# Women Look at Biology
# Looking at Women

Edited by Ruth Hubbard,
Mary Sue Henifin and Barbara Fried
with the collaboration of Vicki Druss
and Susan Leigh Star

# Women Look at Biology Looking at Women

## A Collection of Feminist Critiques

G. K. Hall & Co.    Boston, Massachusetts

Schenkman Publishing Co.    Cambridge, Massachusetts

PERMISSION HAS BEEN GRANTED TO REPRINT THE FOLLOWING MATERIAL: *Datha Clapper Brack, "Displaced—The Midwife by the Male Physician," from* Women and Health, *Vol. 1, No 6 (November/December 1976), courtesy of* Women and Health, *SUNY, College at Old Westbury. Emily Dickinson, "The Brain—is wider than the Sky—," from* The Poems of Emily Dickinson, *ed. Thomas H. Johnson (Cambridge, Mass:. Harvard University Press), copyright 1951, 1955 by the President and Fellows of Harvard College, by permission of the publisher and the Trustees of Amherst College. Susan Griffin, excerpt from* Voices, *courtesy of the author. "Mid-Point," from* Women, *Vol. 4, No. 4, Pt. 1 (1976), courtesy of the author, copyright 1976 by* Women: A Journal of Liberation. *Lanayre Liggera, "Invocation," from* All Our Lives: A Women's Songbook *(Oakland, California: Diana Press), copyright by the author. Karen Lindsay, "Falling off the Roof," from* Falling off the Roof *(Cambridge, Mass.: Alice James Books, 1976), courtesy of the author and Alice James Books. Morton-Norwich Products, Inc., "Norforms" advertisement. June Namias, "Cycles," courtesy of the author. Yoko Ono, "Fish," from* Grapefruit *(New York: Simon and Schuster), copyright 1964, 1970, by Yoko Ono, by permission of Simon and Schuster. Monica Raymond, poem, and Judy Greenberg, drawing, "Anorexia," from* RT, *Vol 5, No. 2 (April-June 1976) courtesy of* State and Mind: People Look at Psychology *(formerly* Rough Times/The Radical Therapist). *Adrienne Rich, "Snapshots of a Daughter-in-Law," from* Snapshots of a Daughter-in-Law, Poems, 1954–1962, *by permission of W.W. Norton and Co., copyright 1956, 1957, 1958, 1959, 1960, 1961, 1962, 1963, 1967 by Adrienne Rich Conrad.*

*Library of Congress Cataloging in Publication Data*

Main entry under title:

Women look at biology looking at women.

    Bibliography: p. 213
    1. Women—Biology—Physiology—Philosophy.  2. Feminism.  3. Sexism.
I. Hubbard, Ruth, 1924–  II. Henifin, Mary Sue.  III. Fried, Barbara, 1951–
QP34.5.W65  1979    301.41'2      79-1445
ISBN 0-8161-9003-3   cloth
ISBN 87073-896-8   paper

*This publication is printed on permanent/durable acid-free paper*
MANUFACTURED IN THE UNITED STATES OF AMERICA

*To the many women, past and present, who have constricted their aspirations to fit within what they were told were the limitations of their biology.*

*The Brain—is wider than the Sky—*
*For—put them side by side—*
*The one the other will contain*
*With ease—and You—beside—*

*The Brain is deeper than the sea—*
*For—hold them—Blue to Blue—*
*The one the other will absorb—*
*As Sponges—Buckets—do—*

*The Brain is just the weight of God—*
*For—Heft them—Pound for Pound—*
*And they will differ—if they do—*
*As Syllable from Sound—*

*—Emily Dickinson, c. 1862*

# Contents

# List of Figures

# Preface and Acknowledgments

This book originated from our participation—one as teacher, two as students—in a seminar on Biology and Women's Issues held at Radcliffe College in the fall of 1975. An important and initially surprising finding was the lack of adequate information on women's biology. Apparently, the wrong questions have been asked, inappropriate methods have been used, and answers have been value-laden.

Recognizing the ramifications of this, we came to understand that women's biology not only is not destiny, but is often not even biology. Our biology was not created by God the Father, but by his human sons. As a result, it contains a number of convenient myths that bolster sexist social practices. We decided to communicate some of our discoveries in this collection of essays, some written by seminar participants and others by colleagues subsequently contacted. The book concludes with a lengthy bibliography. Although this listing is far from complete, it provides the resources necessary to initiate explorations into a variety of areas in biology and medicine which are of concern to women.

This book has been more than three years in the making, in a rapidly growing area of study and activity. The references in some articles therefore do not include the most recent publications. However the bibliography includes references through the beginning of 1979.

The inspiration for this book came from all the seminar participants. Susan Leigh Star and Vicki Druss were part of the initial editorial group, but other demands necessitated their moving too far away to sustain the daily working relationship that this effort has required.

When we first decided to publish a collection on "Biology and Women's Issues," we selected as its title a modification of the first line of Emily Dickinson's poem, "The Brain is Wider than the Sky...," because it conveys our feelings about our human capacity to think, to dream, to describe the world. Although it was later impressed on us that such a title was too vague for our purposes, we want to share the poem with our readers. Therefore, we begin the book with it as an epigraph and thank the Harvard University Press for permission to reprint it.

We also wish to thank both the Committee on Instruction of Currier House at Radcliffe College and the Harvard Committee on General Education for sponsoring the seminar, and the staff of Currier House for their hospitality and support. We are grateful for financial assistance from the office of Dr. Matina Horner, President of Radcliffe College; the Radcliffe Union of Students; Radcliffe Education for Action; and the Milton Fund of Harvard University. Much help and advice has come from our publishers. Many friends and colleagues have helped each and all of us; although they remain unnamed, they are not forgotten.

*Cambridge, Massachusetts*
*September, 1978*

# Introduction

*". . . what is a woman? I assure you, I do not know. I do not believe
that you know. I do not believe that anyone can know until she has
expressed herself in all the arts and professions open to human skill."*
*—Virginia Woolf, "Professions for Women"*

Can you find what's wrong with this description?

> The chief distinction in the intellectual powers of the two sexes
> is shown by man's attaining to a higher eminence, in whatever
> he takes up, than can woman—whether requiring deep thought,
> reason, or imagination, or merely the use of the senses and hands.
> If two lists were made of the most eminent men and women in
> poetry, painting, sculpture, music (inclusive both of composition
> and performance), history, science, and philosophy, with half-
> a-dozen names under each subject, the two lists would not bear
> comparison. We may also infer . . . that if men are capable of a
> decided preeminence over women in many subjects, the average
> of mental power in man must be above that of woman. . . . [1]

In modern literary jargon, the problem might be described as Catch-22.
Among feminists, it is termed "androcentric (i.e., male-centered) bias."
In a court of law, one might call it "double jeopardy," since women
are tried twice for the offense of being born female: at birth when
assigned to a life of mental ineptitude, and at death when the lives thus
lived are judged inept. In biological circles, however, this "problem"
is sometimes attributed to sexual selection, a concept formulated by
Charles Darwin that has become a cornerstone of modern evolutionary
theory.

While many people have long acknowledged an unavoidable sub-
jectivity in perceptions of reality, most continue to attribute objectivity
to scientists and science. But science is the result of a process in which
nature is filtered through a coarse-meshed sieve; only items that
scientists consider worthy of notice are retained. Since scientists are a

rather small group of people—mostly economically and socially privileged, university-educated Caucasians, and predominantly male—there is every reason to assume that, like other human productions, science reflects the outlook and interests of its producers.

Scientists do not ask all possible questions that are amenable to their methodology: only those arousing their curiosity and interest, or the curiosity and interest of supporting organizations. They do not accept all possible answers: only those congruent with the implicit assumptions that form the basis of their understanding of the world (an understanding shared with most of their "educated" contemporaries). Furthermore, the very methodology of science limits its applicability to repeatable and measureable phenomena. This discounts vast areas of human experience, indeed most facets of our relationships with our-

*Figure 1.  Staff of the Department of Chemistry, Massachusetts Institute of Technology (1899–1900), including Ellen Swallow Richards (1842–1911), instructor of the first biology course at M.I.T. (M.I.T. Historical Collections)*

selves, fellow humans, other living beings, and the inanimate world. When scientific knowledge is held superior to other ways of knowing, it serves to devalue or invalidate much of people's daily experience.

Recent critiques have exposed ideological biases in psychology, anthropology, and other social sciences, documenting the ways in which these disciplines often serve the narrow interests of disciples and social peer groups.[2] However, the myth of scientific objectivity has minimized similar criticism of the natural sciences. The time has come to evaluate the "interesting" questions in biology, and to ask why androcentric scientists find them interesting. For instance, among billions of animal species, why have certain ones been studied repeatedly and in great detail, while others have been ignored? Until very recently, greater attention has been focused on the social structure among Savannah baboons than on chimpanzees or gibbons. Could this be because it has been easy to stereotype baboon social behavior as hierarchical, with relatively rigid sex roles? Could it be because chimpanzees have very fluid relationships with one another that are difficult to stereotype by sex except for the fact that females nourish the unweaned young?

Is it an accident that among billions of insect species, those whose social behavior easily conforms to rigid roles are the ones that have caught the imaginations of naturalists from the nineteenth century onward? The "scientific" language of the last century is still in use—ant and bee societies still contain slaves and queens, as well as workers and soldiers. Yet we hear almost nothing about the behavior of insects whose social arrangements do not lend themselves to analogies reinforcing human social arrangements that many people think of as "natural."

Turning to studies of our own species, is it an accident that scientists have been primarily interested in exploring contraceptive techniques that tamper with the *female* reproductive system, following the curious logic that because "fertility in women depends upon so many finely balanced factors . . . it should be easy to interfere with the process at many different stages . . . ?"[3] Would it not be more sensible to conclude that it is more difficult and riskier to tamper with a woman's reproductive system than a man's *because* the woman's system is made up of "so many finely balanced factors?"

These examples suggest a few of the ways in which the objects selected for scientific study and the manner of study are used to reinforce the interests or preconceptions of the studiers. There is clearly not enough time or money for scientists to ask all possible questions. So one asks only those that promise to lead somewhere. The question is *who gets to decide where?*

This book is our beginning to an answer. Each essay stands alone.

Read together, one detects certain recurring themes. The first of them concerns how we use language, or more accurately, how language uses us.

For a long time, biologists, anthropologists, psychologists, and physicians have allocated a great deal of attention to the question, *why are males masculine and females feminine?* They have searched busily for *the* answer: just the right ratio of nature to nurture required to produce our present schizophrenic state of affairs. Nature, it appears, works in ways mysterious and widely varying, depending upon which man of science we accept. Darwin thought that nature enabled us to distinguish boys from girls by endowing the latter with inferior traits at birth—a difference established through sexual selection. John Money contends that nature controls behavior through our hormones and the differing spatial orientation of our genitals (also a favorite with psychologist Erik Erikson). Others have located nature's imprint in a differential development of the left and right sides of our brains. And the sociobiologists locate it in our genes, which supposedly endow women with an updated version of the "mothering instinct"—the evolutionary adaptation that equips women to guard parental investment in our offspring.

While proclaiming erudite and sometimes contradictory answers, they have drawn attention away from the absurdity of the original question:

*Why are males masculine and females feminine?*

Why? Because that is what we call them. If we began to call "prototypic" females Amazons, before long we would have by definition a race of Amazons. A useful tool, if one gets to pick the names. Having designated the disease of "femininity," we can easily prove its existence by innumerable symptoms of "feminine" behavior which females display to be worthy of their given name. When men display similar behavior, the bimodal model is preserved by saying that they, too, can occasionally be "feminine" (persistence, however, indicates "abnormality").

The fact is that people vary a great deal: they come in a wide range of sizes, shapes, tempers, and talents. We can find examples to support almost any bimodal system we choose to construct: fat people are good humored, thin people, nervous; short people are aggressive, tall people, happy-go-lucky; poor people are shiftless, rich people, circumspect; blacks (jews) are warm, whites (gentiles) reserved. As Humpty-Dumpty says, the important thing is who has power to choose the names.[4]

The limits of our language present the limits of "reality" as we know it. Scientists control our thought not only by their choice of subjects but in their manner of description. Much scientific writing employs the passive voice, a ponderous, authority-laden style that carries an automatic sanctification of the subject under discussion. As linguist Julia Stanley points out, when B. F. Skinner writes that "The punishment of sexual behavior changes sexual behavior," he makes passive the statement "*someone* is choosing to punish and thereby change the sexual behavior of certain people."[5] Sexual behavior is punishable. Women are reinforceable. The ecosphere is manipulable. The missing questions that this linguistic structure conceals are: *by whom? for what purpose? in whose interest? under what conditions?*

Looking at scientific language, we must provide answers to these questions and delegitimatize language that masks authority and authorship. Behind every statement that something is doable are a doer and a motive.

The development of a highly technical language has always been defended by scientists as necessary for "precision." Whether this is true or not, it is important to look at the *social* function of technical language. Learning what scientists are talking about requires a long period of apprenticeship, a molding of one's consciousness to fit the information into a precise, *publicly inaccessible* mode.

Scientific knowledge must be maximally accessible and minimally obfuscated through language. Language is power in the scientific mode as elsewhere; if we want the power that comes with information to be more widely shared, we must dispense with the elite protection of unnecessary complexities (including nomenclature) and challenge them wherever we find them.

There is another pitfall that has been largely ignored, or at least underrated. It is called by all sorts of jargon such as "experimenter expectancy," but what it boils down to is that more often than not, we find what we look for. Indeed, one can prove almost any hypothesis if one gets to set the terms of the experiment: to choose appropriate conditions, ask appropriate questions, select appropriate controls. And if one does a thorough job, the conclusion will have that quality of obviousness that scientists so enjoy at the end of meticulous research. And it really *is* obvious, for it fits what we believe about the world; but the reason it fits so well is that it is founded on those very beliefs. Thus were discovered the four humors of the body, leeching to cure disease, the inheritance of acquired characteristics, and countless cast-off theories with which we engender feelings of superiority in present-day biology students. But so also were discovered the accepted theories we examine in this book—they and many others.

Most self-fulfilling theories are devised without intent to defraud, and when they are debunked, they—at worst—damage the reputations of their authors. When such theories become effective tools for oppression, however, they are social dangers. So for example, if some scientists who believe (wish?) that women's mental lives are controlled by the physical demands of their reproductive systems (or that blacks are intellectually inferior) proceed to "prove" these hypotheses by devising the necessary tests, asking the right questions, finding appropriate subjects, and then come to the obvious conclusions, sexism (or racism) becomes part of the scientific dogma.

For two centuries men have avoided the ethics of the "woman question," as they have avoided issues of racial oppression, by claiming to base the relevant political decisions on laws of nature. "It would no doubt be a great boon to the human race," men might say "if women *could* do the marvelous things men have done, like vote, own property, and get a proper education. But such was clearly not nature's design when she gave females a 'head almost too small for intellect but just big enough for love.' "[6] If women had been meant to vote, we would not have been born with a uterus.

That the division by sex of power and privilege in society is politically motivated and not based on biology is sufficiently obvious to have been said often by many different people. But in practically every generation, there arise new prophets of "biology as destiny," and each is fêted as a new Galileo who must be protected against political persecution—this time from enraged women rather than from the church.

The new science of sociobiology would have us believe that women stay home with the children because their eggs are large (hence metabolically more expensive) than their husbands' sperm and that women's "nurturing instinct" has evolved to guard these biological "investments." Though the message to women has been somewhat altered in the century since Darwin, it remains intact: we reproduce, therefore we are. And if our reproductive functions are no longer our sole destiny, they certainly remain our most sanctioned calling. Almost fifty years ago, Virginia Woolf noted Mr. John Langdon Davies' warning that "when children cease to be altogether desirable, women cease to be altogether necessary."[7] From all appearances, androcentric scientists still remain comfortable in that conviction.

What can we do to restructure and rename our scientific world? Twenty years ago a similar question was presented to playwright Lorraine Hansberry about the black revolution in this country. She said the answer is simple: one uses everything one has. One fights in the schools, at the polls, in the marketplace, in the streets. "And, in the process, [we] must have no regard whatsoever for labels and pursed lips in the light of [our] efforts. The acceptance of our present condition is the only form of extremism which discredits us before our children."[8]

And so it is for us, concerned about the impact the science of biology continues to have on our lives. We fight with science's own tools, refuting illogical and self-serving explanations, exposing unsubstantiated claims, disclaiming poorly conceived and inadequately controlled experiments. We fight by turning our talents and money to overcome what are medical problems for some women: to try to cure dysmenorrhea rather than "prove" that it makes us unfit executives, to find safe forms of contraception rather than "prove" how much safer the pill is than death by hanging. And we fight by helping women into positions of responsibility in all facets of science, within the traditional professions and outside them, as health workers, researchers, administrators, policy makers, as teachers of future generations of scientists and as their students.

This book is our contribution. We hope that it will plant seeds in the minds of many women; that it will provoke many of us to ask pointed and incisive (read "unfeminine") questions and to insist on finding honest answers. Only then can we split our present, limited, empirical world wide open and expose fracture faces that we did not even know were concealed within it. Or, to conclude with words of Mary Wollstonecraft, written nearly two hundred years ago,

> Let [women's] faculties have room to unfold,
> and their virtues to gain strength, and then
> determine where the whole sex must stand in the
> intellectual scale.

*September, 1977*

## Notes

1. Charles Darwin, *The Origin of Species and The Descent of Man* (New York: Modern Library Edition), pp. 873–874.
2. *See* for examples M. Kay Martin and Barbara Voorhis, *Female of the Species* (New York: Columbia University Press, 1975); Rayna R. Reiter, ed., *Toward an Anthropology of Women,* (New York: Monthly Review Press, 1975); Naomi Weisstein, "Psychology Constructs the Female," *Woman in Sexist Society,* Vivian Gornick and Barbara Moran, eds. (New York: Basic Books, 1971), pp. 207–224; and Phyllis Chesler, *Women and Madness* (Garden City, NY: Doubleday, 1972).
3. Clive Wood, *Birth Control: Now and Tomorrow* (London: Peter Davies, 1969), pp. 36–37.
4. The actual quotation is:
   "When *I* use a word," Humpty Dumpty said, in rather a scornful tone, "it means exactly what I choose it to mean—neither more nor less."/ "The question is," said Alice, "whether you *can* make words mean so many different things."/ "The question is," said Humpty Dumpty, "which is to be master—that's all."
   Lewis Carroll, *Through the Looking Glass,* ed. Martin Gardner (New American Library) p. 269.
5. Julia P. Stanley, "Nominalized Passives" (Paper delivered at the Linguistic Society of America, Chapel Hill, NC, July 1972).
6. C.D. Meigs, "Lecture on Some of the Distinctive Characteristics of the Female" (Paper delivered at the Jefferson Medical College, Philadelphia, PA, 1847), p. 67; *See* Mary Roth Walsh's essay, "The Quirls of a Woman's Brain."
7. Virginia Woolf, *A Room of One's Own* (New York: Harcourt, Brace and World, 1957), p. 116.
8. Lorraine Hansberry, "To Be Young, Gifted and Black," *Her Own Words,* adapted by Robert Nemiroff (Englewood Cliffs, NJ: Prentice-Hall, 1969), pp. 213–214.

# Part One:

## What Is a Woman?

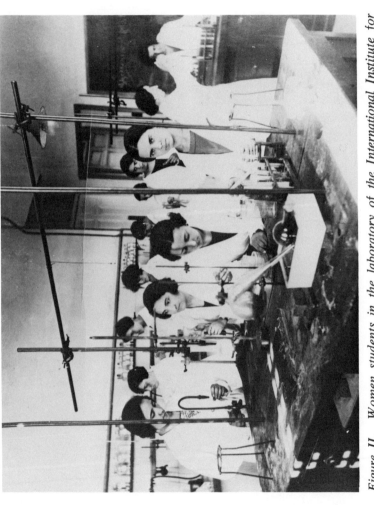

*Figure II. Women students in the laboratory of the International Institute for girls in Spain (1920). The laboratory was founded by Mary Louise Foster (1865–1960), Associate Professor of chemistry at Smith College (1908–1933). (M.I.T. Historical Collections)*

# Introduction

Following David Copperfield's wise example, we begin at the beginning, with the question so fascinating to scientists of the past century: What *is* a woman?

After years of speculation and research, there is very little we can add to what our ancestors observed millenia ago, without the aid of rats, baboons, electrodes and personal interviews. Most women can menstruate, become pregnant, and breast-feed their babies. Most men, for their part, can inseminate women, thereby contributing half the genetic material of the next generation.

To these bare bones, each society has fitted its own notions about behaviors appropriate to each sex. Anthropologists have shown that these notions can be diametrically opposed in different societies; what may be considered fit only for the goose in one, will be the sole province of the gander in another.[1] This fact, however, has hindered few from proclaiming their particular assignment of roles as natural, innate, and commendable. From the moment of birth, each of us is admitted to a social club whose membership, at least until the recent advent of transsexual surgery, has been considered fixed for life. The rules of this membership are often the most stringent that will ever be invoked to govern our conduct, as the Miss Peach Cartoon painfully reminds us:[2]

What name we will be called, what will be the color of the first article
of clothing hung on our still unconscious bodies, what toys we will
play with, what we will be taught in school (indeed whether we go to
school), what books we will read, what our life's work will be, how
much (if at all) we will be paid for it—no aspect of our lives has seemed
too large or too small to be subject to sexual differentiation.

No wonder, then, that we need rarely resort to physical examination
to determine the sex of an individual. Society provides us with clues
more readily detected, as Edwin Lewis neatly illustrates:

> A four year old who had visited a family in which there was a
> new baby was later asked at home whether the baby was a boy
> or a girl. "I don't know," she replied. "It's so hard to tell at that
> age, especially with the clothes off."[3]

The fact is, we can never see each other with our "societal clothes"
off. Rather, scientists have offered us a reversed "emperor's new clothes"
by proclaiming that they can undress the emperor. *Homo sapiens,* they
tell us, stands splendidly naked before us if only we carefully observe
the behavior of rats, monkeys, apes, and peahens, as though these
animals were humans stripped of enculturation. Scientists have compared humans across history and continents, thinking to discard all
our varying, societal "clothes" as acquired characteristics, while establishing the remaining as innate. They have tried to isolate "pure"
behavior—that not subject to environmental influence—by measuring
brain waves, or observing pre-language infants. And they have tried
to cancel out interference from "impure" behavior by drawing their
subjects from what they perceive to be identical environments.

So far each approach has proved inadequate to the herculean task—
rather like poking around the ruins of a great fire to find the match
that started it all. We cannot regulate human environments as we do the
life of a laboratory rat; we can match up quantifiable statistics, but we
can't measure the quality of a person's world. Often correspondences
we think we see between ourselves and other species are invented by us.
When we study our history, other cultures, infants, or possibly even
our brain waves, we are looking at phenomena which are themselves
products of sexually dimorphic societies. We are looking at these
phenomena with eyes accustomed to find, perhaps even hoping to find,
sexual dimorphism in everything we see.

"The story of our lives becomes our lives,"[4] writes Adrienne Rich;
a truth perhaps nowhere more apt than in the story of our lives as
women or men. We live in a world which for so long has reported our
differences as essential, that our lives have come to assume the shape of

this profound conviction. How, then, are we to disentangle the two, how do we isolate nature from nurture? How can we expect to discover what is responsible for each of the differences we observe between the sexes? And why do we care?

It is difficult to understand the investment so many people have in believing sex differences to be profound, and biologically based, unless one realizes that the ideology of sex differences serves a social function. Like other forms of biological determinism, it can be used to reinforce the *status quo* by implying that what is, must be. In a scientifically oriented, politically liberal society like ours, existing inequalities between the sexes can no longer be derived from Laws of God or Man. Hence there is every motivation to shore them up with appropriate Laws of Nature.

The three essays in this section on evolution, sex and gender, and brain asymmetry show how our preconceptions inform efforts to examine scientifically the evolution of the human species and the personal development of individuals. They show that scientific "facts," like all others, are generated within a social context; and that their context pushes certain realizations into the foreground, while others readily merge with the background of the unnoticed and unremarked, and hence are undescribed.[5] As women begin to pull forth facts which have been previously ignored, while pushing back others which have received more notoriety than their substance merits, one thing becomes clear: not only must we not believe that our biology is our destiny; we must reexamine whether it is even our biology.

What then is a woman? The question is still happily unresolved, and surely will remain so for quite some time. The only accurate answer we are likely to find will take years to reveal itself, and will only begin to do so when women's lives are lived unhampered by the myths of what a woman is not.

*August, 1977*

## Notes

1. *See* for example, Margaret Mead, *Male and Female: A Study of the Sexes in a Changing World* (New York: Dell, 1949); Ernestine Friedl, *Women and Men: An Anthropologist's View* (New York: Holt, Rinehart, Winston, 1975); and C. J. Matthiasson, ed. *Many Sisters: Women in Cross-cultural Perspective,* (New York: Free Press, 1974.)
2. Mell Lazarus, "Miss Peach" (New York: Field Enterprises, 1970.)
3. Edwin C. Lewis, *Developing Woman's Potential* (Ames, IA: Iowa State University Press, 1968), p. 16.
4. Adrienne Rich, *Twenty One Love Poems* (Emery, CA: Effie's Press, 1976). No. 18.
5. For a more general discussion of the operation of unconscious "foregrounding" and "backgrounding" *see* Mary Douglas's Introduction to *Implicit Meanings* (London: Routledge and Kegan Paul, 1975). The social context in which scientists operate has a strong effect on their often quite unconscious choices of what they notice and what they relegate to the background. To quote Mary Douglas, "Most forms of social life call for coherence and clear definition. The same energy that constrains disruptive passions and creates a certain pattern of society also organises knowledge in a compatible workable useable form."

# Ruth Hubbard

# Have Only Men Evolved?

*". . . with the dawn of scientific investigation it might have been hoped that the prejudices resulting from lower conditions of human society would disappear, and that in their stead would be set forth not only facts, but deductions from facts, better suited to the dawn of an intellectual age . . . .*

*The ability, however, to collect facts, and the power to generalize and draw conclusions from them, avail little, when brought into direct opposition to deeply rooted prejudices."*
—Eliza Burt Gamble, *The Evolution of Woman* (1894)

Science is made by people who live at a specific time in a specific place and whose thought patterns reflect the truths that are accepted by the wider society. Because scientific explanations have repeatedly run counter to the beliefs held dear by some powerful segments of the society (organized religion, for example, has its own explanations of how nature works), scientists are sometimes portrayed as lone heroes swimming against the social stream. Charles Darwin (1809–82) and his theories of evolution and human descent are frequently used to illustrate this point. But Darwinism, on the contrary, has wide areas of congruence with the social and political ideology of nineteenth-century Britain and with Victorian precepts of morality, particularly as regards the relationships between the sexes. And the same Victorian notions still dominate contemporary biological thinking about sex differences and sex roles.

## Science and the Social Construction of Reality

For humans, language plays a major role in generating reality. Without words to objectify and categorize our sensations and place them in relation to one another, we cannot evolve a tradition of what is real in the world. Our past experience is organized through language into our

*Figure III. Reconstruction of Neanderthal "household"*
*(American Museum of Natural History)*

history within which we have set up new verbal categories that allow us to assimilate present and future experiences. If every time we had a sensation we gave it a new name, the names would have no meaning: lacking consistency, they could not arrange our experience into reality. For words to work, they have to be used consistently and in a sufficient variety of situations so that their volume—what they contain and exclude —becomes clear to all their users.

If I ask a young child, "Are you hungry?", she must learn through experience that "yes" can produce a piece of bread, a banana, an egg, or an entire meal; whereas "yes" in answer to "Do you want orange juice?" always produces a tart, orange liquid.

However, all acts of naming happen against a backdrop of what is socially accepted as real. The question is *who* has social sanction to define the larger reality into which one's everyday experiences must fit in order that one be reckoned sane and responsible. In the past, the Church had this right, but it is less looked to today as a generator of new definitions of reality, though it is allowed to stick by its old ones even when they conflict with currently accepted realities (as in the case of miracles). The State also defines some aspects of reality and can generate what George Orwell called Newspeak in order to interpret the world for its own political purposes. But, for the most part, at present science is the most respectable legitimator of new realities.

However, what is often ignored is that science does more than merely define reality; by setting up first the definitions—for example, three-dimensional (Euclidian) space—and then specific relationships within them—for example, parallel lines never meet—it automatically renders suspect the sense experiences that contradict the definitions. If we want to be respectable inhabitants of the Euclidian world, every time we see railroad tracks meet in the distance we must "explain" how what we are seeing is consistent with the accepted definition of reality. Furthermore, through society's and our personal histories, we acquire an investment in our sense of reality that makes us eager to enlighten our children or uneducated "savages," who insist on believing that railroad tracks meet in the distance and part like curtains as they walk down them. (Here, too, we make an exception for the followers of some accepted religions, for we do not argue with equal vehemence against our fundamentalist neighbors, if they insist on believing literally that the Red Sea parted for the Israelites, or that Jesus walked on the Sea of Galilee.)

Every theory is a self-fulfilling prophecy that orders experience into the framework it provides. Therefore, it should be no surprise that almost any theory, however absurd it may seem to some, has its supporters. The mythology of science holds that scientific theories lead to the truth because they operate by consensus: they can be tested by different scientists, making their own hypotheses and designing independent experiments to test them. Thus, it is said that even if one or another scientists "misinterprets" his or her observations, the need for consensus will weed out fantasies and lead to reality. But things do not work that way. Scientists do not think and work independently. Their "own" hypotheses ordinarily are formulated within a context of theory, so that their interpretations by and large are sub-sets within the prevailing orthodoxy. Agreement therefore is built into the process and need tell us little or nothing about "truth" or "reality." Of course, scientists often disagree, but their quarrels usually are about details that do not contradict fundamental beliefs, whichever way they are resolved.[1] To overturn orthodoxy is no easier in science than in philosophy, religion, economics, or any of the other disciplines through which we try to comprehend the world and the society in which we live.

The very language that translates sense perceptions into scientific reality generates that reality by lumping certain perceptions together and sorting or highlighting others. But what we notice and how we describe it depends to a great extent on our histories, roles, and expectations as individuals and as members of our society. Therefore, as we move from the relatively impersonal observations in astronomy, physics and chemistry into biology and the social sciences, our science is increasingly affected by the ways in which our personal and social experience determine what we are able or willing to perceive as real about ourselves and the organisms around us. This is not to accuse scientists of being deluded or dishonest, but merely to point out that, like other people, they find it difficult to see the social biases that are built into the very fabric of what they deem real. That is why, by and large, only children notice that the emperor is naked. But only the rare child hangs on to that insight; most of them soon learn to see the beauty and elegance of his clothes.

In trying to construct a coherent, self-consistent picture of the world, scientists come up with questions and answers that depend on their perceptions of what has been, is, will be, and can be. There is no such thing as objective, value-free science. An era's science is part of its politics, economics and sociology: it is generated by them and in turn helps to generate them. Our personal and social histories mold what we perceive to be our biology and history as organisms, just as our biology plays its part in our social behavior and perceptions. As scientists, we learn to examine the ways in which our experimental methods can bias our answers, but we are not taught to be equally wary of the biases introduced by our implicit, unstated and often unconscious beliefs about the nature of reality. To become conscious of these is more difficult than anything else we do. But difficult as it may seem, we must try to do it if our picture of the world is to be more than a reflection of various aspects of ourselves and of our social arrangements.[2]

## Darwin's Evolutionary Theory

It is interesting that the idea that Darwin was swimming against the stream of accepted social dogma has prevailed, in spite of the fact that many historians have shown his thinking fitted squarely into the historical and social perspective of his time. Darwin so clearly and admittedly was drawing together strands that had been developing over long periods of time that the questions why he was the one to produce the synthesis and why it happened just then have clamored for answers. Therefore, the social origins of the Darwinian synthesis have been probed by numerous scientists and historians.

A belief that all living forms are related and that there also are deep connections between the living and non-living has existed through much of recorded human history. Through the animism of tribal cultures that endows everyone and everything with a common spirit; through more elaborate expressions of the unity of living forms in some Far Eastern and Native American belief systems; and through Aristotelian notions of connectedness runs the theme of one web of life that includes humans among its many strands. The Judaeo-Christian world view has been exceptional—and I would say flawed—in setting man (and I mean the male of the species) apart from the rest of nature by making him the namer and ruler of all life. The biblical myth of the creation gave rise to the separate and unchanging species which that second Adam, Linnaeus (1707-78), later named and classified. But even Linnaeus—though he began by accepting the belief that all existing species had been created by Jehovah during that one week long ago ("Nulla species nova")—had his doubts about their immutability by the time he had identified more than four thousand of them: some species appeared to be closely related, others seemed clearly transitional. Yet as Eiseley has pointed out, it is important to realize that:

> Until the scientific idea of 'species' acquired form and distinctness there could be no dogma of 'special' creation in the modern sense. This form and distinctness it did not possess until the naturalists of the seventeenth century began to substitute exactness of definition for the previous vague characterizations of the objects of nature.[3]

And he continues:

> ... it was Linnaeus with his proclamation that species were
> absolutely fixed since the beginning who intensified the theological
> trend. ... Science, in its desire for classification and order, ...
> found itself satisfactorily allied with a Christian dogma whose
> refinements it had contributed to produce.

Did species exist before they were invented by scientists with their
predilection for classification and naming? And did the new science, by
concentrating on differences which could be used to tell things apart,
devalue the similarities that tie them together? Certainly the Linnaean
system succeeded in congealing into a relatively static form what had
been a more fluid and graded world that allowed for change and hence
for a measure of historicity.

The hundred years that separate Linnaeus from Darwin saw the
development of historical geology by Lyell (1797-1875) and an incipient
effort to fit the increasing number of fossils that were being uncovered
into the earth's newly discovered history. By the time Darwin came
along, it was clear to many people that the earth and its creatures had
histories. There were fossil series of snails; some fossils were known
to be very old, yet looked for all the world like present-day forms; others
had no like descendants and had become extinct. Lamarck (1744-1829),
who like Linnaeus began by believing in the fixity of species, by 1800
had formulated a theory of evolution that involved a slow historical
process, which he assumed to have taken a very, very long time.

Possibly one reason the theory of evolution arose in Western, rather
than Eastern, science was that the descriptions of fossil and living forms
showing so many close relationships made the orthodox biblical view
of the special creation of each and every species untenable; and the
question, how living forms merged into one another, pressed for an
answer. The Eastern philosophies that accepted connectedness and
relatedness as givens did not need to confront this question with the
same urgency. In other words, where evidences of evolutionary change
did not raise fundamental contradictions and questions, evolutionary
theory did not need to be invented to reconcile and answer them. However one, and perhaps the most, important difference between Western
evolutionary thinking and Eastern ideas of organismic unity lies in
the materialistic and historical elements, which are the earmark of
Western evolutionism as formulated by Darwin.

Though most of the elements of Darwinian evolutionary theory
existed for at least hundred years before Darwin, he knit them into a

consistent theory that was in line with the mainstream thinking of his time. Irvine writes:

> The similar fortunes of liberalism and natural selection are significant. Darwin's matter was as English as his method. Terrestrial history turned out to be strangely like Victorian history writ large. Bertrand Russell and others have remarked that Darwin's theory was mainly 'an extension to the animal and vegetable world of laissez faire economics.' As a matter of fact, the economic conceptions of utility, pressure of population, marginal fertility, barriers in restraint of trade, the division of labor, progress and adjustment by competition, and the spread of technological improvements can all be paralleled in *The Origin of Species*. But so, alas, can some of the doctrines of English political conservatism. In revealing the importance of time and the hereditary past, in emphasizing the persistence of vestigial structures, the minuteness of variations and the slowness of evolution, Darwin was adding Hooker and Burke to Bentham and Adam Smith. The constitution of the universe exhibited many of the virtues of the English constitution.[4]

One of the first to comment on this congruence was Karl Marx (1818-83) who wrote to Friedrich Engels (1820-95) in 1862, three years after the publication of *The Origin of Species*:

> It is remarkable how Darwin recognizes among beasts and plants his English society with its division of labour, competition, opening up of new markets, 'inventions,' and the Malthusian 'struggle for existence.' It is Hobbes's 'bellum omnium contra omnes,' [war of all against all] and one is reminded of Hegel's *Phenomenology,* where civil society is described as a 'spiritual animal kingdom,' while in Darwin the animal kingdom figures as civil society.[5]

A similar passage appears in a letter by Engels:

> The whole Darwinist teaching of the struggle for existence is simply a transference from society to living nature of Hobbes's doctrine of 'bellum omnium contra omnes' and of the bourgeois-economic doctrine of competition together with Malthus's theory of population. When this conjurer's trick has been performed ... the same theories are transferred back again from organic nature into history and now it is claimed that their validity as eternal laws of human society has been proved.[5]

The very fact that essentially the same mechanism of evolution through natural selection was postulated independently and at about the same time by two English naturalists, Darwin and Alfred Russel Wallace (1823-1913), shows that the basic ideas were in the air—which is not to deny that it took genius to give them logical and convincing form.

Darwin's theory of *The Origin of Species by Means of Natural Selection,* published in 1859, accepted the fact of evolution and undertook to explain how it could have come about. He had amassed large quantities of data to show that historical change had taken place, both from the fossil record and from his observations as a naturalist on the Beagle. He pondered why some forms had become extinct and others had survived to generate new and different forms. The watchword of evolution seemed to be: be fruitful and modify, one that bore a striking resemblance to the ways of animal and plant breeders. Darwin corresponded with many breeders and himself began to breed pigeons. He was impressed by the way in which breeders, through careful selection, could use even minor variations to elicit major differences, and was searching for the analog in nature to the breeders' techniques of selecting favorable variants. A prepared mind therefore encountered Malthus's *Essay on the Principles of Population* (1798). In his *Autobiography,* Darwin writes:

> In October 1838, that is, fifteen months after I had begun my systematic enquiry, I happened to read for amusement Malthus on *Population,* and being well prepared to appreciate the struggle for existence which everywhere goes on from long-continued observation of the habits of animals and plants, it at once struck me that under these circumstances favourable variations would tend to be preserved and unfavourable ones to be destroyed. The result of this would be the formation of new species. Here, then, I had at last got a theory by which to work.[6]

Incidentally, Wallace also acknowledged being led to his theory by reading Malthus. Wrote Wallace:

> The most interesting coincidence in the matter, I think, is, that I, *as well as Darwin,* was led to the theory itself through Malthus. . . . It suddenly flashed upon me that all animals are necessarily thus kept down—'the struggle for existence'—while *variations,* on which I was always thinking, must necessarily often be *beneficial,* and would then cause those varieties to increase while the injurious variations diminished.[7] (Wallace's italics)

Both, therefore, saw in Malthus's struggle for existence the working of a natural law which effected what Herbert Spencer had called the "survival of the fittest."

The three principal ingredients of Darwin's theory of evolution are: endless variation, natural selection from among the variants, and the resulting survival of the fittest. Given the looseness of many of his arguments—he credited himself with being an expert wriggler—it is surprising that his explanation has found such wide acceptance. One reason probably lies in the fact that Darwin's theory was historical and materialistic, characteristics that are esteemed as virtues; another, perhaps in its intrinsic optimism—its notion of progressive development of species, one from another—which fit well into the meritocratic ideology encouraged by the early successes of British mercantilism, industrial capitalism and imperialism.

But not only did Darwin's interpretation of the history of life on earth fit in well with the social doctrines of nineteenth-century liberalism and individualism. It was used in turn to support them by rendering them aspects of natural law. Herbert Spencer is usually credited with having brought Darwinism into social theory. The body of ideas came to be known as social Darwinism and gained wide acceptance in Britain and the United States in the latter part of the nineteenth and on into the twentieth century. For example, John D. Rockefeller proclaimed in a Sunday school address:

> The growth of a large business is merely the survival of the fittest .... The American Beauty rose can be produced in the splendor and fragrance which bring cheer to its beholder only by sacrificing the early buds which grow up around it. This is not an evil tendency in business. It is merely the working-out of a law of nature and a law of God.[8]

The circle was therefore complete: Darwin consciously borrowed from social theorists such as Malthus and Spencer some of the basic concepts of evolutionary theory. Spencer and others promptly used Darwinism to reinforce these very social theories and in the process bestowed upon them the force of natural law.[9]

## Sexual Selection

It is essential to expand the foregoing analysis of the mutual influences of Darwinism and nineteenth-century social doctrine by looking critically at the Victorian picture Darwin painted of the relations between the sexes, and of the roles that males and females play in the evolution of animals and humans. For although the ethnocentric bias of Darwinism is widely acknowledged, its blatant sexism—or more correctly, androcentrism (male-centeredness)—is rarely mentioned, presumably because it has not been noticed by Darwin scholars, who have mostly been men. Already in the nineteenth century, indeed within Darwin's life time, feminists such as Antoinette Brown Blackwell and Eliza Burt Gamble called attention to the obvious male bias pervading his arguments.[10,11] But these women did not have Darwin's or Spencer's professional status or scientific experience; nor indeed could they, given their limited opportunities for education, travel and participation in the affairs of the world. Their books were hardly acknowledged or discussed by professionals, and they have been, till now, merely ignored and excluded from the record. However, it is important to expose Darwin's androcentrism, and not only for historical reasons, but because it remains an integral and unquestioned part of contemporary biological theories.

Early in *The Origin of Species,* Darwin defines sexual selection as one mechanism by which evolution operates. The Victorian and androcentric biases are obvious:

> This form of selection depends, not on a struggle for existence in relation to other organic beings or to external conditions, but on a struggle of individuals of one sex, generally males, for the possession of the other sex.[12]

And,

> Generally, the most vigorous males, those which are best fitted for their places in nature, will leave most progeny. But in many cases, victory depends not so much on general vigor, as on having special weapons confined to the male sex.

The Victorian picture of the active male and the passive female becomes even more explicit later in the same paragraph:

> the males of certain hymenopterous insects [bees, wasps, ants] have been frequently seen by that inimitable observer, M. Fabre, fighting for a particular female who sits by, an apparently unconcerned beholder of the struggle, and then retires with the conqueror.

Darwin's anthropomorphizing continues, as it develops that many male birds "perform strange antics before the females, which, standing by as spectators, at last choose the most attractive partner." However, he worries that whereas this might be a reasonable way to explain the behavior of peahens and female birds of paradise whose consorts anyone can admire, "it is doubtful whether [the tuft of hair on the breast of the wild turkey-cock] can be ornamental in the eyes of the female bird." Hence Darwin ends this brief discussion by saying that he "would not wish to attribute all sexual differences to this agency."

Some might argue in defense of Darwin that bees (or birds, or what have you) do act that way. But the very language Darwin uses to describe these behaviors disqualifies him as an "objective" observer. His animals are cast into roles from a Victorian script. And whereas no one can claim to have solved the important methodological question of how to disembarrass oneself of one's anthropocentric and cultural biases when observing animal behavior, surely one must begin by trying.

After the publication of *The Origin of Species,* Darwin continued to think about sexual selection, and in 1871, he published *The Descent of Man and Selection in Relation to Sex,* a book in which he describes in much more detail how sexual selection operates in the evolution of animals and humans.

In the aftermath of the outcry *The Descent* raised among fundamentalists, much has been made of the fact that Darwin threatened the special place Man was assigned by the Bible and treated him as though he was just another kind of animal. But he did nothing of the sort. The Darwinian synthesis did not end anthropocentrism or androcentrism in biology. On the contrary, Darwin made them part of biology by presenting as "facts of nature" interpretations of animal behavior that reflect the social and moral outlook of his time.

In a sense, anthropocentrism is implicit in the fact that we humans have named, catalogued, and categorized the world around us, including ourselves. Whether we stress our upright stance, our opposable thumbs, our brain, or our language, to ourselves we are creatures apart and very different from all others. But the scientific view of ourselves is also profoundly androcentric. *The Descent of Man* is quite literally *his* journey. Elaine Morgan rightly says:

> It's just as hard for man to break the habit of thinking of himself as central to the species as it was to break the habit of thinking of himself as central to the universe. He sees himself quite unconsciously as the main line of evolution, with a female satellite revolving around him as the moon revolves around the earth. This not only causes him to overlook valuable clues to our ancestry, but sometimes leads him into making statements that are arrant and demonstrable nonsense .... Most of the books

forget about [females] for most of the time. They drag her on
stage rather suddenly for the obligatory chapter on Sex and Reproduction, and then say: 'All right, love, you can go now,' while
they get on with the real meaty stuff about the Mighty Hunter
with his lovely new weapons and his lovely new straight legs
racing across the Pleistocene plains. Any modifications of her
morphology are taken to be imitations of the Hunter's evolution,
or else designed solely for his delectation.[13]

To expose the Victorian roots of post-Darwinian thinking about
human evolution, we must start by looking at Darwin's ideas about
sexual selection in *The Descent,* where he begins the chapter entitled
"Principles of Sexual Selection" by setting the stage for the active,
pursuing male:

With animals which have their sexes separated, the males necessarily differ from the females in their organs of reproduction; and
these are the primary sexual characters. But the sexes differ in
what Hunter has called secondary sexual characters, which are
not directly connected with the act of reproduction; for instance,
the male possesses certain organs of sense or locomotion, of
which the female is quite destitute, or has them more highly-
developed, in order that he may readily find or reach her; or
again the male has special organs of prehension for holding
her securely.[14]

Moreover, we soon learn:

in order that the males should seek efficiently, it would be necessary
that they should be endowed with strong passions; and the acquire-
ment of such passions would naturally follow from the more
eager leaving a larger number of offspring than the less eager.[15]

But Darwin is worried because among some animals, males and females
do not appear to be all that different:

a double process of selection has been carried on; that the males
have selected the more attractive females, and the latter the more
attractive males . . . . But from what we know of the habits of
animals, this view is hardly probable, for the male is generally
eager to pair with any female.[16]

Make no mistake, wherever you look among animals, eagerly promiscuous males are pursuing females, who peer from behind languidly
drooping eyelids to discern the strongest and handsomest. Does it not

sound like the wishfulfillment dream of a proper Victorian gentleman? This is not the place to discuss Darwin's long treatise in detail. Therefore, let this brief look at animals suffice as background for his section on Sexual Selection in Relation to Man. Again we can start on the first page: "Man is more courageous, pugnacious and energetic than woman, and has more inventive genius."[17] Among "savages," fierce, bold men are constantly battling each other for the possession of women and this has affected the secondary sexual characteristics of both. Darwin grants that there is some disagreement whether there are "inherent differences" between men and women, but suggests that by analogy with lower animals it is "at least probable." In fact, "Woman seems to differ from man in mental disposition, chiefly in her greater tenderness and less selfishness,"[18] for:

> Man is the rival of other men; he delights in competition, and this leads to ambition which passes too easily into selfishness. These latter qualities seem to be his natural and unfortunate birthright.

This might make it seem as though women are better than men after all, but not so:

> The chief distinction in the intellectual powers of the two sexes is shown by man's attaining to a higher eminence, in whatever he takes up, than can women—whether requiring deep thought, reason, or imagination, or merely the use of the senses and hands. If two lists were made of the most eminent men and women in poetry, painting, sculpture, music (inclusive both of composition and performance), history, science, and philosophy, with half-a-dozen names under each subject, the two lists would not bear comparison. We may also infer . . . that if men are capable of a decided pre-eminence over women in many subjects, the average of mental power in man must be above that of woman. . . . [Men have had] to defend their females, as well as their young, from enemies of all kinds, and to hunt for their joint subsistence. But to avoid enemies or to attack them with success, to capture wild animals, and to fashion weapons, requires the aid of the higher mental faculties, namely, observation, reason, invention, or imagination. These various faculties will thus have been continually put to the test and selected during manhood.[19]

"Thus," the discussion ends, "man has ultimately become superior to woman" and it is a good thing that men pass on their characteristics to their daughters as well as to their sons, "otherwise it is probable that man would have become as superior in mental endowment to woman, as the peacock is in ornamental plumage to the peahen."

So here it is in a nutshell: men's mental and physical qualities were constantly improved through competition for women and hunting, while women's minds would have become vestigial if it were not for the fortunate circumstance that in each generation daughters inherit brains from their fathers.

Another example of Darwin's acceptance of the conventional mores of his time is his interpretation of the evolution of marriage and monogamy:

> ... it seems probable that the habit of marriage, in any strict sense of the word, has been gradually developed; and that almost promiscuous or very loose intercourse was once very common throughout the world. Nevertheless, from the strength of the feeling of jealousy all through the animal kingdom, as well as from the analogy of lower animals ... I cannot believe that absolutely promiscuous intercourse prevailed in times past. . . .[20]

Note the moralistic tone; and how does Darwin know that strong feelings of jealousy exist "all through the animal kingdom?" For comparison, it is interesting to look at Engels, who working largely from the same early anthropological sources as Darwin, had this to say:

> As our whole presentation has shown, the progress which manifests itself in these successive forms [from group marriage to pairing marriage to what he refers to as "monogamy supplemented by adultery and prostitution"] is connected with the peculiarity that women, but not men, are increasingly deprived of the sexual freedom of group marriage. In fact, for men group marriage actually still exists even to this day. What for the woman is a crime entailing grave legal and social consequences is considered honorable in a man or, at the worse, a slight moral blemish which he cheerfully bears . . . . Monogamy arose from the concentration of considerable wealth in the hands of a single individual—a man—and from the need to bequeath this wealth to the children of that man and of no other. For this purpose, the monogamy of the woman was required, not that of the man, so this monogamy of the woman did not in any way interfere with open or concealed polygamy on the part of the man.[21]

Clearly, Engels did not accept the Victorian code of behavior as our natural biological heritage.

## Sociobiology: A New Scientific Sexism

The theory of sexual selection went into a decline during the first half of this century, as efforts to verify some of Darwin's examples showed that many of the features he had thought were related to success in mating could not be legitimately regarded in that way. But it has lately regained its respectability, and contemporary discussions of reproductive fitness often cite examples of sexual selection.[22] Therefore, before we go on to discuss human evolution, it is helpful to look at contemporary views of sexual selection and sex roles among animals (and even plants).

Let us start with a lowly alga that one might think impossible to stereotype by sex. Wolfgang Wickler, an ethologist at the University of Munich, writes in his book on sexual behavior patterns (a topic which Konrad Lorenz tells us in the Introduction is crucial in deciding which sexual behaviors to consider healthy and which diseased):

> Even among very simple organisms such as algae, which have threadlike rows of cells one behind the other, one can observe that during copulation the cells of one thread act as males with regard to the cells of a second thread, but as females with regard to the cells of a third thread. The mark of male behavior is that the cell actively crawls or swims over to the other; the female cell remains passive.[23]

The circle is simple to construct: one starts with the Victorian stereotype of the active male and the passive female, then looks at animals, algae, bacteria, people, and calls all passive behavior feminine, active or goal-oriented behavior masculine. And it works! The Victorian stereotype is biologically determined: even algae behave that way.

But let us see what Wickler has to say about Rocky Mountain Bighorn sheep, in which the sexes cannot be distinguished on sight. He finds it "curious":

> that between the extremes of rams over eight years old and lambs less than a year old one finds every possible transition in age, but no other differences whatever; the bodily form, the structure of the horns, and the color of the coat are the same for both sexes.

Now note: ". . . the typical female behavior is absent from this pattern." Typical of what? Obviously not of Bighorn sheep. In fact we are told that "even the males often cannot recognize a female," indeed, "the females are only of interest to the males during rutting season." How does he know that the males do *not* recognize the females? Maybe these

sheep are so weird that most of the time they relate to a female as
though she were just another sheep, and whistle at her (my free translation of "taking an interest") only when it is a question of mating.
But let us get at last to how the *females* behave. That is astonishing,
for it turns out:

> that *both* sexes play two roles, either that of the male or that of
> the young male. Outside the rutting season the females behave
> like young males, during the rutting season like aggressive older
> males. (Wickler's italics)

In fact:

> There is a line of development leading from the lamb to the high
> ranking ram, and the female animals ( ♀ ) behave exactly as
> though they were in fact males ( ♂ ) whose development was re-
> tarded . . . . We can say that the only fully developed mountain
> sheep are the powerful rams. . . .

At last the androcentric paradigm is out in the open: females are always
measured against the standard of the male. Sometimes they are like
young males, sometimes like older ones; but never do they reach what
Wickler calls "the final stage of fully mature physical structure and
behavior possible to this species." That, in his view, is reserved for
the rams.

Wickler bases this discussion on observations by Valerius Geist,
whose book, *Mountain Sheep,* contains many examples of how androcentric biases can color observations as well as interpretations and
restrict the imagination to stereotypes. One of the most interesting is
the following:

> Matched rams, usually strangers, begin to treat each other like
> females and clash until one acts like a female. This is the loser
> in the fight. The rams confront each other with displays, kick
> each other, threat jump, and clash till one turns and accepts the
> kicks, displays, and occasional mounts of the larger without
> aggressive displays. The loser is not chased away. The point of
> the fight is not to kill, maim, or even drive the rival off, but to
> treat him like a female.[24]

This description would be quite different if the interaction were
interpreted as something other than a fight, say as a homosexual en-
counter, a game, or a ritual dance. The fact is that it contains none of
the elements that we commonly associate with fighting. Yet because Geist

casts it into the imagery of heterosexuality and aggression, it becomes perplexing.

There would be no reason to discuss these examples if their treatments of sex differences or of male/female behavior were exceptional. But they are in the mainstream of contemporary sociobiology, ethology, and evolutionary biology.

A book that has become a standard reference is George Williams's *Sex and Evolution.*[25] It abounds in blatantly biased statements that describe as "careful" and "enlightened" research reports that support the androcentric paradigm, and as questionable or erroneous those that contradict it. Masculinity and femininity are discussed with reference to the behavior of pipefish and seahorses; and cichlids and catfish are judged downright abnormal because both sexes guard the young. For present purposes it is sufficient to discuss a few points that are raised in the chapter entitled "Why Are Males Masculine and Females Feminine and, Occasionally, Vice-Versa?"

The very title gives one pause, for if the words masculine and feminine do not mean of, or pertaining, respectively, to males and females, what *do* they mean—particularly in a scientific context? So let us read.

On the first page we find:

> Males of the more familiar higher animals take less of an interest in the young. In courtship they take a more active role, are less discriminating in choice of mates, more inclined toward promiscuity and polygamy, and more contentious among themselves.

We are back with Darwin. The data are flimsy as ever, but doesn't it sound like a description of the families on your block?

The important question is who are these "more familiar higher animals?" Is their behavior typical, or are we familiar with them because, for over a century, androcentric biologists have paid disproportionate attention to animals whose behavior resembles those human social traits that they would like to interpret as biologically determined and hence out of our control?

Williams' generalization quoted above gives rise to the paradox that becomes his chief theoretical problem:

> Why, if each individual is maximizing its own genetic survival should the female be less anxious to have her eggs fertilized than a male is to fertilize them, and why should the young be of greater interest to one than to the other?

Let me translate this sentence for the benefit of those unfamiliar with current evolutionary theory. The first point is that an individual's *fitness* is measured by the number of her or his offspring that survive to reproductive age. The phrase, "the survival of the fittest," therefore signifies the fact that evolutionary history is the sum of the stories of those who leave the greatest numbers of descendants. What is meant by each individual "maximizing its own genetic survival" is that every one tries to leave as many viable offspring as possible. (Note the implication of conscious intent. Such intent is not exhibited by the increasing number of humans who intentionally *limit* the numbers of their offspring. Nor is one, of course, justified in ascribing it to other animals.)

One might therefore think that in animals in which each parent contributes half of each offspring's genes, females and males would exert themselves equally to maximize the number of offspring. However, we know that according to the patriarchal paradigm, males are active in courtship, whereas females wait passively. This is what Williams means by females being "less anxious" to procreate than males. And of course we also know that "normally" females have a disproportionate share in the care of their young.

So why these asymmetries? The explanation: "The *essential* difference between the sexes is that females produce large immobile gametes and males produce small mobile ones" (my italics). This is what determines their "different optimal strategies." So if you have wondered why men are promiscuous and women faithfully stay home and care for the babies, the reason is that males "can quickly replace wasted gametes and be ready for another mate," whereas females "can not so readily replace a mass of yolky eggs or find a substitute father for an expected litter." Therefore females must "show a much greater degree of caution" in the choice of a mate than males.

E. O. Wilson says the same thing somewhat differently:

> One gamete, the egg, is relatively very large and sessile; the other, the sperm, is small and motile. . . . The egg possesses the yolk required to launch the embryo into an advanced state of development. Because it represents a considerable energetic investment on the part of the mother the embryo is often sequestered and protected, and sometimes its care is extended into the postnatal period. *This is the reason why* parental care is *normally* provided by the female. . . .[26] (my italics)

Though these descriptions fit only some of the animal species that reproduce sexually, and are rapidly ceasing to fit human domestic arrangements in many portions of the globe,[27] they do fit the patriarchal model of the household. Clearly, androcentric biology is busy as ever trying to provide biological "reasons" for a particular set of human social arrangements.

The ethnocentrism of this individualistic, capitalistic model of evolutionary biology and sociobiology with its emphasis on competition and "investments," is discussed by Sahlins in his monograph, *The Use and Abuse of Biology*.[5] He gives many examples from other cultures to show how these theories reflect a narrow bias that disqualifies them from masquerading as descriptions of universals in biology. But, like other male critics, Sahlins fails to notice the obvious androcentrism.

About thirty years ago, Ruth Herschberger wrote a delightfully funny book called *Adam's Rib*,[28] in which she spoofed the then current androcentric myths regarding sex differences. When it was reissued in 1970, the book was not out of date. In the chapter entitled "Society Writes Biology," she juxtaposes the then (and now) current patriarchal scenario of the dauntless voyage of the active, agile sperm toward the passively receptive, sessile egg to an improvised "matriarchal" account. In it the large, competent egg plays the central role and we can feel only pity for the many millions of miniscule, fragile sperm most of which are too feeble to make it to fertilization.

This brings me to a question that always puzzles me when I read about the female's larger energetic investment in her egg than the male's in his sperm: there is an enormous disproportion in the *numbers* of eggs and sperms that participate in the act of fertilization. Does it really take more "energy" to generate the one or relatively few eggs than the large excess of sperms required to achieve fertilization? In humans the disproportion is enormous. In her life time, an average woman produces about four hundred eggs, of which in present-day Western countries, she will "invest" only in about 2.2.[29] Meanwhile the average man generates several billions of sperms to secure those same 2.2 investments!

Needless to say, I have no idea how much "energy" is involved in producing, equipping and ejaculating a sperm cell along with the other necessary components of the ejaculum that enable it to fertilize an egg, nor how much is involved in releasing an egg from the ovary, reabsorbing it in the oviduct if unfertilized (a partial dividend on the investment), or incubating 2.2 of them to birth. But neither do those who propound the existence and importance of women's disproportionate energetic investments. Furthermore, I attach no significance to these questions, since I do not believe that the details of our economic and social arrangements reflect our evolutionary history. I am only trying to show how feeble is the "evidence" that is being put forward to argue the evolutionary basis (hence *naturalness*) of woman's role as home-maker.

The recent resurrection of the theory of sexual selection and the ascription of asymmetry to the "parental investments" of males and females are probably not unrelated to the rebirth of the women's movement. We should remember that Darwin's theory of sexual selection was put forward in the midst of the first wave of feminism.[30] It seems that when women threaten to enter as equals into the world of affairs, androcentric scientists rally to point out that our *natural* place is in the home.

## The Evolution of Man

Darwin's sexual stereotypes are doing well also in the contemporary literature on human evolution. This is a field in which facts are few and specimens are separated often by hundreds of thousands of years, so that maximum leeway exists for investigator bias. Almost all the investigators have been men; it should therefore come as no surprise that what has emerged is the familiar picture of Man the Toolmaker.

*Figure IV.   Discussion of the Piltdown skull.*
*(American Museum of Natural History)*

This extends so far that when skull fragments estimated to be 250,000 years old turned up among the stone tools in the gravel beds of the Thames at Swanscombe and paleontologists decided that they are probably those of a female, we read that "The Swanscombe woman, or her husband, was a maker of hand axes . . . ."[31] (Imagine the reverse: The Swanscombe man, or his wife, was a maker of axes . . . .) The implication is that if there were tools, the Swanscombe *woman* could not have made them. But we now know that even apes make tools. Why not women?

Actually, the idea that the making and use of tools were the main driving forces in evolution has been modified since paleontological finds and field observations have shown that apes both use and fashion tools. Now the emphasis is on the human use of tools as weapons for hunting. This brings us to the myth of Man the Hunter, who had to invent not only tools, but also the social organization that allowed him to hunt big animals. He also had to roam great distances and learn to cope with many and varied circumstances. We are told that this entire constellation of factors stimulated the astonishing and relatively rapid development of his brain that came to distinguish Man from his ape cousins. For example, Kenneth Oakley writes:

> Men who made tools of the standard type . . . must have been capable of forming in their minds images of the ends to which they laboured. Human culture in all its diversity is the outcome of this capacity for conceptual thinking, but the leading factors in its development are tradition coupled with invention. The primitive hunter made an implement in a particular fashion largely because as a child he watched his father at work or because he copied the work of a hunter in a neighbouring tribe. The standard hand-axe was not conceived by any one individual *ab initio,* but was the result of exceptional individuals in successive generations not only copying but occasionally improving on the work of their predecessors. As a result of the co-operative hunting, migrations and rudimentary forms of barter, the traditions of different groups of primitive hunters sometimes became blended.[32]

It seems a remarkable feat of clairvoyance to see in such detail what happened some 250,000 years in pre-history, complete with the little boy and his little stone chipping set just like daddy's big one.

It is hard to know what reality lurks behind the reconstructions of Man Evolving. Since the time when we and the apes diverged some fifteen million years ago, the main features of human evolution that one can read from the paleontological finds are the upright stance, reduction in the size of the teeth, and increase in brain size. But finds

*Figure V.   Reconstructions of the "progression of prehistoric man,"*
*including (second from left) the Piltdown hoax.*
*(American Museum of Natural History)*

are few and far between both in space and in time until we reach the
Neanderthals some 70,000 to 40,000 years ago—a jaw or skull, teeth,
pelvic bones, and often only fragments of them.[33] From such bits of
evidence as these come the pictures and statues we have all seen of
that line of increasingly straight and upright, and decreasingly hairy
and ape-like men marching in single file behind *Homo sapiens,* carrying
their clubs, stones, or axes; or that other one of a group of beetle-
browed and bearded hunters bending over the large slain animal they
have brought into camp, while over on the side long-haired, broad-bot-
tomed females nurse infants at their pendulous breasts.

Impelled, I suppose, by recent feminist critiques of the evolution of
Man the Hunter, a few male anthropologists have begun to take note
of Woman the Gatherer, and the stereotyping goes on as before. For
example Howells, who acknowledges these criticisms as just, none-
theless assumes "the classic division of labor between the sexes" and
states as fact that stone age men roamed great distances "on behalf of
the whole economic group, while the women were restricted to within
the radius of a fraction of a day's walk from camp." Needless to say, he
does not *know* any of this.

One can equally well assume that the responsibilities for providing
food and nurturing young were widely dispersed through the group

*Figure VI. Reconstruction of Cro-Magnon "mammoth hunters."*
*(American Museum of Natural History)*

that needed to cooperate and devise many and varied strategies for survival. Nor is it obvious why tasks needed to have been differentiated by sex. It makes sense that the gatherers would have known how to hunt the animals they came across; that the hunters gathered when there was nothing to catch, and that men and women did some of each, though both of them probably did a great deal more gathering than hunting. After all, the important thing was to get the day's food, not to define sex roles. Bearing and tending the young have not necessitated a sedentary way of life among nomadic peoples right to the present, and both gathering and hunting probably required movement over large areas in order to find sufficient food. Hewing close to home probably accompanied the transition to cultivation, which introduced the necessity to stay put for planting, though of course not longer than required to harvest. Without fertilizers and crop rotation, frequent moves were probably essential parts of early farming.

Being sedentary ourselves, we tend to assume that our foreparents heaved a great sigh of relief when they invented agriculture and could at last stop roaming. But there is no reason to believe this. Hunter/gatherers and other people who move with their food still exist. And what has been called the agricultural "revolution" probably took considerably longer than all of recorded history. During this time, presumably some people settled down while others remained nomadic, and some did some of each, depending on place and season.

We have developed a fantastically limited and stereotypic picture of ways of life that evolved over many tens of thousands of years, and no doubt varied in lots of ways that we do not even imagine. It is true that by historic times, which are virtually now in the scale of our evolutionary history, there were agricultural settlements, including a few towns that numbered hundreds and even thousands of inhabitants. By that time labor was to some extent divided by sex, though anthropologists have shown that right to the present, the division can be different in different places. There are economic and social reasons for the various delineations of sex roles. We presume too much when we try to read them in the scant record of our distant prehistoric past.

Nor are we going to learn them by observing our nearest living relatives among the apes and monkeys, as some biologists and anthropologists are trying to do. For one thing, different species of primates vary widely in the extent to which the sexes differ in both their anatomy and their social behavior, so that one can find examples of almost any kind of behavior one is looking for by picking the appropriate animal. For another, most scientists find it convenient to forget that present-day apes and monkeys have had as long an evolutionary history as we have had, since the time we and they went our separate ways many millions of years ago. There is no theoretical reason why their behavior should tell us more about our ancestry than our behavior tells us about theirs. It is only anthropocentrism that can lead someone to magine that "A possible preadaptation to human ranging for food is the behavior of the large apes, whose groups move more freely and widely compared to gibbons and monkeys, and whose social units are looser."[34] But just as in the androcentric paradigm men evolved while women cheered from the bleachers, so in the anthropocentric one, humans evolved while the apes watched from the trees. This view leaves out not only the fact that the apes have been evolving away from us for as long a time as we from them, but that certain aspects of their evolution may have been a response to our own. So, for example, the evolution of human hunting habits may have put a serious crimp into the evolution of the great apes and forced them to stay in the trees or to hurry back into them.

The current literature on human evolution says very little about the role of language, and sometimes even associates the evolution of language with tool use and hunting—two purportedly "masculine" characteristics. But this is very unlikely because the evolution of language probably went with biological changes, such as occurred in the structure of the face, larynx, and brain, all slow processes. Tool use and hunting, on the other hand, are cultural characteristics that can evolve much more quickly. It is likely that the more elaborate use of tools, and the social arrangements that go with hunting and gathering, developed in part as a consequence of the expanded human repertory of capacities and needs that derive from our ability to communicate through language.

It is likely that the evolution of speech has been one of the most powerful forces directing our biological, cultural, and social evolution, and it is surprising that its significance has largely been ignored by biologists. But, of course, it does not fit into the androcentric paradigm. No one has ever claimed that women can not talk; so if men are the vanguard of evolution, humans must have evolved through the stereotypically male behaviors of competition, tool use, and hunting.

## *How to Learn Our History? Some Feminist Strategies*

How *did* we evolve? Most people now believe that we became who we are by a historical process, but, clearly, we do not know its course, and must use more imagination than fact to reconstruct it. The mythology of science asserts that with many different scientists all asking their own questions and evaluating the answers independently, whatever personal bias creeps into their individual answers is cancelled out when the large picture is put together. This might conceivably be so if scientists were women and men from all sorts of different cultural and social backgrounds who came to science with very different ideologies and interests. But since, in fact, they have been predominantly university-trained white males from privileged social backgrounds, the bias has been narrow and the product often reveals more about the investigator than about the subject being researched.

Since women have not figured in the paradigm of evolution, we need to rethink our evolutionary history. There are various ways to do this:

(1) We can construct one or several estrocentric (female-centered) theories. This is Elaine Morgan's approach in her account of *The Descent of Woman* and Evelyn Reed's in *Woman's Evolution*.[35] Except as a way of parodying the male myths, I find it unsatisfactory because it locks the authors into many of the same unwarranted suppositions that underlie those very myths. For example, both accept the view that our behavior is biologically determined, that what we do is a result of what we were or did millions of years ago. This assumption is unwarranted given the enormous range of human adaptability and the rapid rate of human social and cultural evolution. Of course, there is a place for myth-making and I dream of a long poem that sings women's origins

and tells how we felt and what we did; but I do not think that carefully constructed "scientific" mirror images do much to counter the male myths. Present-day women do not know what prehistoric hunter/gatherer women were up to any more than a male paleontologist like Kenneth Oakley knows what the little toolmaker learned from his dad.

(2) Women can sift carefully the few available facts by paring away the mythology and getting as close to the raw data as possible. And we can try to see what, if any, picture emerges that could lead us to questions that perhaps have *not* been asked and that should, and could, be answered. One problem with this approach is that many of the data no longer exist. Every excavation removes the objects from their locale and all we have left is the researchers' descriptions of what they saw. Since we are concerned about unconscious biases, that is worrisome.

(3) Rather than invent our own myths, we can concentrate, as a beginning, on exposing and analyzing the male myths that hide our overwhelming ignorance, "for when a subject is highly controversial—and any question about sex is that—one cannot hope to tell the truth."[36] Women anthropologists have begun to do this. New books are being written, such as *The Female of the Species*[37] and *Toward an Anthropology of Women*,[38] books that expose the Victorian stereotype that runs through the literature of human evolution, and pull together relevant anthropological studies. More important, women who recognize an androcentric myth when they see one and who are able to think beyond it, must do the necessary work in the field, in the laboratories, and in the libraries, and come up with ways of seeing the facts and of interpreting them.

None of this is easy, because women scientists tend to hail from the same socially privileged families and be educated in the same elite universities as our male colleagues. But since we are marginal to the mainstream, we may find it easier than they to watch ourselves push the bus in which we are riding.

As we rethink our history, our social roles, and our options, it is important that we be ever wary of the wide areas of congruence between what are obviously ethno- and androcentric assumptions and what we have been taught are the scientifically proven facts of our biology. Darwin was right when he wrote that "False facts are highly injurious to the progress of science, for they often endure long . . . ."[39] Androcentric science is full of "false facts" that have endured all too long and that serve the interests of those who interpret as women's biological heritage the sexual and social stereotypes we reject. To see our alternatives is essential if we are to acquire the space in which to explore who we are, where we have come from, and where we want to go.

## Notes

I want to thank Gar Allen, Rita Arditti, Steve Gould and my colleagues in the editorial group that has prepared this book for their helpful criticisms of an earlier version of this manuscript.

1. For a discussion of this process, *see* Thomas S. Kuhn, *The Structure of Scientific Revolutions,* 2nd ed. (University of Chicago Press, 1970).

2. Berger and Luckmann have characterized this process as "trying to push a bus in which one is riding." [Peter Berger and Thomas Luckmann, *The Social Construction of Reality* (Garden City: Doubleday & Co., 1966) p. 12.]. I would say that, worse yet, it is like trying to look out of the rear window to *watch* oneself push the bus in which one rides.

3. Loren Eiseley, *Darwin's Century* (Garden City: Doubleday & Co., Anchor Books Edition, 1961), p. 24.

4. William Irvine, *Apes, Angels, and Victorians* (New York: McGraw-Hill, 1972), p. 98.

5. Quoted in Marshall Sahlins, *The Use and Abuse of Biology* (Ann Arbor: University of Michigan Press, 1976), pp. 101–102.

6. Francis Darwin, ed., *The Autobiography of Charles Darwin* (New York: Dover Publications, 1958), pp. 42–43.

7. *Ibid.,* pp. 200–201.

8. Richard Hofstadter, *Social Darwinism in American Thought* (Boston: Beacon Press, 1955), p. 45.

9. Though not himself a publicist for social Darwinism like Spencer, there can be no doubt that Darwin accepted its ideology. For example, near the end of *The Descent of Man* he writes: "There should be open competition for all men; and the most able should not be prevented by laws or customs from succeeding best and rearing the largest number of offspring." Marvin Harris has argued that Darwinism, in fact, should be known as biological Spencerism, rather than Spencerism as social Darwinism. For a discussion of the issue, *pro* and *con, see* Marvin Harris, *The Rise of Anthropological Theory: A History of Theories of Culture* (New York: *Thomas Y. Crowell,* 1968), Ch. 5: Spencerism; and responses by Derek Freeman and others in *Current Anthropology* 15 (1974), 211–237.

10. Antoinette Brown Blackwell, *The Sexes Throughout Nature* (New York: G. P. Putnam's Sons, 1975; reprinted Westport, Conn.: Hyperion Press, 1978). Excerpts in which Blackwell argues against Darwin and Spencer have been reprinted in Alice S. Rossi, ed., *The Feminist Papers* (New York: Bantam Books, 1974), pp. 356–377.

11. Eliza Burt Gamble, *The Evolution of Woman: An Inquiry into the Dogma of her Inferiority to Man* (New York: G. P. Putnam's Sons, 1894).

12. Charles Darwin, *The Origin of Species and the Descent of Man* (New York: Modern Library Edition), p. 69.
13. Elaine Morgan, *The Descent of Woman* (New York: Bantam Books, 1973), pp. 3–4.
14. Darwin, *Origin of Species*, p. 567.
15. *Ibid.*, p. 580.
16. *Ibid.*, p. 582.
17. *Ibid.*, p. 867.
18. *Ibid.*, p. 873.
19. *Ibid.*, pp. 873–874.
20. *Ibid.*, p. 895.
21. Frederick Engels, *The Origin of the Family, Private Property and the State*, E. B. Leacock, ed. (New York: International Publishers, 1972), p. 138.
22. One of the most explicit contemporary examples of this literature is E. O. Wilson's *Sociobiology: The New Synthesis* (Cambridge: Harvard University Press, Belknap Press, 1975); *see* especially chapters 1, 14–16 and 27.
23. Wolfgang Wickler, *The Sexual Code: The Social Behavior of Animals and Men* (Garden City: Doubleday, Anchor Books, 1973), p. 23.
24. Valerius Geist, *Mountain Sheep* (Chicago: University of Chicago Press, 1971) p. 190.
25. George C. Williams, *Sex and Evolution* (Princeton: Princeton University Press, 1975).
26. Edward O. Wilson, *Sociobiology: The New Synthesis* (Cambridge: Harvard University Press, Belknap Press, 1975), pp. 316–317. Wilson and others claim that the growth of a mammalian fetus inside its mother's womb represents an energetic "investment" on her part, but it is not clear to me why they believe that. Presumably the mother eats and metabolizes, and some of the food she eats goes into building the growing embryo. Why does that represent an investment of *her* energies? I can see that the embryo of an undernourished woman perhaps requires such an investment—in which case what one would have to do is see that the mother gets enough to eat. But what "energy" does a properly nourished woman "invest" in her embryo (or, indeed, in her egg)? It would seem that the notion of pregnancy as "investment" derives from the interpretation of pregnancy as a debilitating disease that Datha Brack discusses in her essay in this collection.
27. For example, at present in the United States, 24 per cent of households are headed by women and 46 per cent of women work outside the home. The fraction of women who work away from home while raising children is considerably larger in several European countries and in China.

28. Ruth Herschberger, *Adam's Rib* (1948; reprinted ed., New York: Harper and Row, 1970).

29. Furthermore, a woman's eggs are laid down while she is an embryo, hence at the expense of her mother's "metabolic investment." This raises the question whether grandmothers devote more time to grandchildren they have by their daughters than to those they have by their sons. I hope sociobiologists will look into this.

30. Nineteenth-century feminism is often dated from the publication in 1792 of Mary Wollstonecraft's (1759–1797) *A Vindication of the Rights of Woman;* it continued right through Darwin's century. Darwin was well into his work at the time of the Seneca Falls Declaration (1848), which begins with the interesting words:

> When, in the course of human events, it becomes necessary for one portion of the family of man to assume among the people of the earth a position different from that which they have hitherto occupied, but one to which the *laws of nature and of nature's God* entitle them . . . (my italics).

And John Stuart Mill (1806–1873) published his essay on *The Subjection of Women* in 1869, ten years after Darwin's *Origin of Species* and two years before the *Descent of Man and Selection in Relation to Sex.*

31. William Howells, *Evolution of the Genus* Homo (Reading: Addison-Wesley Publishing Co., 1973), p. 88.

32. Kenneth P. Oakley, *Man the Toolmaker* (London: British Museum, 1972), p. 81.

33. There are also occasional more perfect skeletons, such as that of *Homo erectus* at Choukoutien, commonly known as Peking Man, who was in fact a woman.

34. Howells, p. 133.

35. Evelyn Reed, *Woman's Evolution* (New York: Pathfinder Press, 1975).

36. Virginia Woolf, *A Room of One's Own* (1945; reprinted ed., Penguin Books, 1970), p. 6.

37. M. Kay Martin and Barbara Voorhis, *Female of the Species* (New York: Columbia University Press, 1975).

38. Rayna R. Reiter, ed., *Toward an Anthropology of Women* (New York: Monthly Review Press, 1975).

39. Darwin, *Origin of Species,* p. 909.

[Ruth Hubbard *is Professor of Biology at Harvard University where she teaches courses dealing with the interaction of science and society. In her scientific research she has shown how the visual pigment molecules in the retina of the eye change when they absorb light and has tried to understand how their changes initiate the signals that travel to the brain. In recent years she has been thinking, writing, and lecturing about how the assumptions scientists make about the world influence their work and how the society in which they live influences their assumptions. She is particularly interested in the way gender—the fact of one's having grown up female or male— affects these questions. She is also interested in health care, particularly as it relates to women. She has written numerous articles and reviews on all these subjects.*]

# Barbara Fried

## Boys Will Be Boys Will Be Boys: The Language of Sex and Gender

Our first record of the use of the word "tomboy" is in 1553 as a term of censure for "a rude, boisterous or forward boy." Twenty-five years later, the censure is redirected to a "girl who behaves like a spirited or boisterous boy," and there it has remained ever since.

While the specific attributes of "tomboyism" have seen some modifications over the past four centuries to conform to our changing notions of impropriety, the function of the label remains the same: to separate from their "normal" sisters those girls who display behavior considered fitting only for boys. It is a word particularly laden with society's fears and disapproval. It is not a word one would expect to find employed as a scientific measure of behavior.

But employed it is. In their celebrated book, *Man and Woman, Boy and Girl,* published in 1972, John Money and Anke Ehrhardt once more enlisted its services to designate deviance: this time of girls who have been exposed, while still in their mothers' wombs, to a greater than usual amount of "male" hormone, an androgen (andros = male, Greek). The list of tomboyish characteristics they compile is more methodical, less poetical, than those of their literary counterparts who found their way into the *Oxford English Dictionary*. It overlooks some of the activities previously condemned so stoutly, while including some which the Elizabethans never would have dreamed of. But "tomboy" conveys still the original distress at discovering girls whose behavior violates the bounds of what is deemed "natural."

Words like tomboy, with all their peculiar distresses and presumptions, have set the terms for every "scientific" exploration of the relationship between sex and gender identity.[1] And it is no longer possible to think we can study that relationship without also studying the role language plays in establishing it—language, that is, in its broadest sense, meaning our collective memory as well as the specific enlistment of that memory in oral and written communication.

Language is the means by which we abstract an experience from its

*Figure VII. Investigators at the Marine Biological Laboratory, Woods Hole, Massachusetts (1896).*
*(Society for the Preservation of New England Antiquities; Baldwin Coolidge, Photograph)*

immediate physical context, and thereby attach to the experience a meaning independent of that context. With the acquisition of language, a child's experience is moved from a purely *subjective* plane to a partially *objective* one. But both these words have implications opposite from those commonly understood. The process of objectification, rather than removing experience from the realm of personal prejudice, injects that prejudice into it for the first time. The child, one would suppose, formerly had no preconceptions through which to understand experience, except perhaps a predisposition to distinguish between pleasure and pain, and a predilection for the former. But the memory newly acquired in language will henceforth prejudice the child's perception of each new experience by setting up the increasingly complex framework in which s/he assigns it meaning.

Throughout our adult life, our accumulated language thus mediates between us and our sensory experience, and then mediates among all our personal syntheses to produce communication. The problems with which we are confronted in any "scientific" study are: To what extent does that medium become the message? And to what extent are the properties we perceive in what we study more accurately attributed to the properties of our own perception? This problem is usually ignored by scientists, who tend to regard language as a neutral medium that can push subjective experience into an objective realm without tampering with its meaning. My particular concern is the extent to which this blithe disregard for the complex interrelationships of language, thought, and reality has underlain (and hence undermined) the work done thus far on sex and gender differentiation.

It is accepted fact that "gender identity is fundamentally established in more or less the same period as native language—in the first two years of life" (*MWBG,* 164). This simultaneous advent of language and gender identity is not coincidental. Language does not simply communicate the link between one's sex and one's gender identity; it *constitutes* that link. The first time a man watched a woman drive a car into a lamp post and observed, "Ah yes, women drivers," he did not merely voice a preexisting correspondence between female genitals and poor driving skills. He created that correspondence. And whether at the time it was, in fact, statistically accurate rapidly becomes unimportant: once such a correspondence is planted in people's minds, they will be predisposed to notice all the examples that bear it out (and a random distribution of driving skills guarantees abundant examples of women with poor ones). And the more examples observed, and hence the more firmly planted the correspondence, the more likely it is to be borne out by statistics as women's behavior changes to conform to their newly acquired identity as automotive incompetents. When the process has gone on long enough, the correspondence is absorbed into our consciousness as an inchoate law of human nature, permanently codified and transmitted to our children in simple units of meaning like "women drivers."

This simple example should make it obvious how complex a task it necessarily must be to isolate the genetic from the environmental sources of sexually differentiated behavior. But scientists have approached the relationship of sex to gender not as a complex system but as a simple dichotomy—whatever isn't sex is gender, and vice versa. In doing so, they have disregarded those mechanisms in biological and cultural evolution that have produced the current state of sexual dimorphism we see before us as "immutable fact." They have disregarded as well those mechanisms in us that influence the way we perceive those "facts." By ignoring these complexities, much of their work founders on the fundamental fallacy of using a language that has become sexually separate and unequal to determine the parameters of sexual separateness and inequality in society—a society whose development is channeled by the same (separate and unequal) language.

The work that John Money and his school have done on sex and gender, which is presented for the non-specialist in the book *Man & Woman, Boy & Girl,* has been accepted by many as the authoritative work in the field. Their methodology can be called into serious question on grounds that are outside the scope of this essay;[2] the grounds to be considered here are the roles language plays in 1) structuring the "facts" that have been studied; 2) structuring the way those facts are perceived; and 3) controlling how those perceptions are communicated to the reader.

## How Language Shapes Reality

The first problem, the role of language in constructing the reality that is being studied, concerns the fallacy of animal analogy (or why mounting behavior in *rats* tells us absolutely nothing about the sexual life of *humans*). Most "sex and gender" research is designed to isolate genetically-controlled sexually dimorphic behavior from that which is culturally induced.[3] There are two ways that behavioral scientists have traditionally tried to achieve this isolation. The first is through exhaustive cross-cultural studies, assuming that whatever common elements of behavior emerge in widely disparate societies must be genetic in origin.[4] The second is to study a "pure" population, one that has not yet defined itself differentially by gender, which usually means pre-verbal infants. Money and Ehrhardt use neither approach, aside from a brief cross-cultural foray into sex-dimorphic erotic behavior. While it is true that both approaches present problems of their own,[5] the alternative methodology that the authors adopt is even more questionable. They have taken as their subjects the rare cases of spontaneously occurring hermaphroditism or sexual ambiguity—persons born with external genital characteristics that do not accord with their chromosomal sex. They have matched these subjects with "normal" controls who share all but those characteristics (that is, they accord in economic class, family situation, IQ, etc.), presuming that whatever *behavioral* differences are observed between the two groups are attributable to the differing *biological* traits.

The following is an example of their methodology in action. Two of the groups they studied were composed of girls who were exposed to excess androgen during their mothers' pregnancies. The first group (ten girls) experienced the increased androgen level as an unknown side effect of the drug progestin, which had been given to their pregnant mothers. Ironically, progestins are "female" hormones, administered to women who were threatened by miscarriages to allow them to carry their pregnancies to term. However, once in the body, progestins were in some instances metabolized to androgens that masculinized *in utero* the external genitals of the girl babies.[6]

The second group (fifteen girls) were afflicted with adrenogenital syndrome, a defect in the adrenal gland that makes it produce greater than normal amounts of androgen. This can now be treated after birth, but not while the infants are still in the womb. Such girls are born with abnormally large clitorises, so large that they are sometimes mistaken for boys.

The fifteen "andrenogenital" girls whom Money and Ehrhardt studied had all been treated soon after birth by reducing their androgen levels to normal and correcting surgically their enlarged clitorises. The ten progestin-affected girls underwent corrective surgery where necessary, and required no further treatment since the influence of progestin on their hormone levels ceased at birth.

Therefore, although all twenty-five girls had been "fetally andro-genized," their biographies were considered otherwise normal: they had two X chromosomes (like all females), their genitals after early surgery were of normal female appearance, their androgen levels after birth were normal, and all of them were raised as females. The authors compared these girls with a normal "control" group matched by age, IQ, socio-economic backround and race. They assumed that whatever behavioral differences were found in the "fetally androgenized" girls could be attributed to a single divergent element in their biographies: that the excess of "prenatal androgens may have left a presumptive effect on the brain, and hence on subsequent behavior" (*MWBG*, 98).

If Money and Ehrhardt were studying rats, their presumption perhaps might be reasonable, but with humans it is not. A high fetal androgen level is *not* the only divergent element in the biographies of these girls. Every one of the families knew that their daughters had been born "abnormal" because they were prenatally androgenized; all the girls were exposed to continuous and extensive treatment by doctors and psychologists; in addition to immediate surgical correction for genitals that were masculine in appearance, some of the girls required vaginal surgery in their teens (surgery Money and Ehrhardt dismiss as minor [97]). And all the girls with adrenogenital syndrome required regular treatment from birth to keep their androgen levels down.

Clearly, the families as well as the girls themselves were aware of their anomalous sexual histories. But the authors never question what effect this knowledge may have had on the parents' and the girls' expectations of their behavior, and hence on that behavior itself. Yet common sense would suggest that the effect could have been great enough to account for what the authors have labeled as the girls' "tomboyish" behavior without contriving to explain it by some hypothetical, ill-defined process in which their brains were "masculinized" *in utero*.

The same disregard of human consciousness as a variable underlies the authors' extensive discussion of dimorphic erotic behavior in humans. They maintain that the sexual fantasies of both women and men universally center on the man's active desire for the woman. This requires that a woman respond to erotic images by "identifying herself with the female to whom men respond [with the result that] she herself becomes the sexual object," while men respond by objectifying the female (252).[7] Even if Money and Ehrhardt *had* documented the universal occurrence of these two different patterns of response to erotic imagery, *which they do not*,[8] they disregard the fact that our language, through every public medium (TV, movies, books, newspapers, etc.), devotes more energy to teaching girls and boys "appropriate" erotic responses than to any other single pursuit. Even if one *were* to prove those patterns to be quite general, what would one prove beyond the fact that one of society's blitzkriegs had had its intended effect? But Money and Ehrhardt would like us to believe they have proved far more:

> Of course, it is possible for a woman as well as a man to respond
> to the visual stimulus of a lover or potential partner in sex . . . .

Her imagery of arousal will, however, tend to be different, as though geared to the premise that he, the man, will be coitally useless to her, except that the stimulus of her receptive body is capable of erecting his penis. Her arousal fantasy tends to build itself around the sentiment and romance of his reacting to her and wanting her—of wanting to hold, caress, and kiss her. If he wants her enough, then his penis will erect and want her too, and perhaps not only once, but again and again for a lasting love affair, and even an entire lifetime. [251]

One can hardly imagine a more concise and illuminating picture of what the male fantasizes to be a woman's inner life. This may be fascinating biography or socio-cultural history, but is it science? And again:

Depictions of sexual intercourse, especially in a movie, are erotically stimulating to women, as well as men, but the same basic difference of identification versus objectification applies once again. The woman viewer is likely to build her erotic excitement into a fantasy of enlarging her repertory of sexual skills, learning something from the female in the movie, with the intention of utilizing it on the next available occasion with her lover or sexual partner. The man viewer builds up a level of erotic excitement, imagining that the woman on the screen is having intercourse with him on the spot. It would not, in fact, be difficult for him to copulate with any live surrogate for the female in the movie. [252]

Certainly there are many women not inclined to dispute the truth of at least the last sentence. But proving something's existence—even (if they had) its wide-spread existence—is not the same thing as proving it is innate.

Through the same faulty methodology, Money and Ehrhardt identify an innate sense of territoriality in males (although they themselves demonstrate in their preceding paragraph how such dimorphic traits as territoriality could arise from divergent childhood experiences):

Territoriality is less prominent in the human male than in various lower species, but some signs of it are evident. Boys rather than girls are youthful explorers, fort-builders, and scouts, and boys are the ones who form gangs or troops that set up territories, dare rivals to trespass, and attack them if they do. [182]

Aside from the fact that what the authors are offering up as biological fact, by an implied analogy to lower animals, is nothing more than undocumented cliché concerning male and female roles in this particular society, any girl who has had the not uncommon experience of being ejected from the fort or banned from the neighborhood troop could tell Money and Ehrhardt that there is nothing very mysterious or genetic about the lack of territoriality she subsequently displays.

But the authors reach that conclusion through the same spurious logic that maintains that Blacks have "natural rhythm." The fallacy goes: "If a particular pattern of behavior recurs frequently enough, we can conclude that it is innate." It is possible to attack this doctrine on a conservative level: that is, by conserving its basic premise but questioning its application. We would then ask, how frequent is "frequently enough?" For example, if research showed that in all but two known human societies women cry more often than men, are we to conclude that the difference is genetic? Or must we conclude on the strength of those two exceptions that the difference is socially acquired? And if the research turns up no exceptions, are we sure that it has considered all available evidence, and without observer bias?[9]

But it is also possible to attack the doctrine on a more radical basis. Even if diligent observations turn up no exceptions to the rule, we may have proved only that the influence of sex-role stereotyping on human history has been more universal than we suspected. Such a possibility should not surprise us, because of the powerful nature of language. It not only shapes our understanding of our past and present; it engenders our future as well, by establishing the limits of our futuristic vision. This is the profound meaning of oppression, that it not only circumscribes our experience, but it also denies us an intuition of that circumscription. It thus guarantees its own perpetuation, because lacking that intuition, we will only wander out of our prescribed limits individually and by accident, when we can be picked off and packed away as deviants one by one. So where the bias is most deep, we should not be surprised to find it universal—but that does not make it any less a bias or any more a biological fact.

This process of implying but never proving a biological basis for an observed behavioral trait underlies much of the research that has been done on sex and gender. Erik Erikson's well-known study on sexual differences in spatial orientation is another example of this phenomenon. Erikson gave 300 children the task of constructing a scene with toys on a table. After studying the resulting configurations, he concluded that girls and boys differ in their utilization of space: girls constructed around what he called "inner space," whereas boys built into "outer space," a difference he predictably ties to their respective anatomies.

Ann Oakley provides an excellent critique of Erikson's study, one that can serve as a model for attacking all similarly constructed experiments:

> Although he does not actually say the differences are innate, the intimate association between them and the body ground-plan means that he is giving a general biological explanation. The words he underlines in the description of the girls' constructions refer to the female's relegation to interior, domestic, sedentary occupations in our society. In the boys' constructions, the underlined words refer to activity, aggression, self-projection, exteriority, and male pursuits. When faced with this result, the question he asks is, 'Does the anatomy of male and female suggest a difference along these lines?' The question he fails to ask is 'How do boys and girls translate this anatomical difference into such detailed and specific differences of activity and interest?' An even more pertinent question is 'Why do these differences so exactly parallel the roles society defines for male and female?' The children in Erikson's sample had had ample time to observe these differences between male and female roles, and ample time to incorporate them into their own play. Aged between ten and twelve, they were far too old to be used for studying the possibility of innate sex differences.

And we can add to Oakley's list of fallacies, one more to be drawn from her earlier description of Erikson's methodology:

> To test whether the qualities he believed he had detected had an objective existence that could be recognized by other observers, he asked other people to sort photographs of the toy scenes into male and female piles, using his criteria: the correlation between his own evaluation and those of the other observers was in fact statistically significant.[10]

## How Language Shapes Our Perceptions of Reality

Erikson's supposed corroboration leads to our second area of concern, the role of language in structuring our perceptions of reality, or the fallacy of the "objective eye witness." "Surely," Erikson seems to argue, "if 150 girls display the same 'female' spatial orientation that mimics their womb, and if all my colleagues detect the same quality of 'femaleness' in their constructions, then there can be no doubt that we have discovered an innate, eternal principle of femaleness." Right?

Wrong. Forty million Frenchmen can be wrong, and they usually are when it comes to their observations on French women. And we should not be surprised to find that they are all wrong in exactly the same way, since they have all "inherited" the same set of cultural prejudices that determine how they see what they look at.

Money and Ehrhardt do not even trouble to call in "disinterested" parties to corroborate their perceptions of their subjects' behavior. They simply arrogate to themselves the role of "objective eye witness" and present their observations as incontrovertible fact. Thus they offer the following "documentation" of tomboyish behavior in the group of fetally androgenized girls:

> The discrepancy between the diagnostic groups and their controls reached statistical significance on the criterion of baby care in the adrenogenital syndrome only. Some girls in this group disliked handling babies and believed they would be awkward and clumsy. By contrast, many of the control girls rated high in enthusiasm for little children; they adored them and took every opportunity to get in close contact with them.

> All control girls were sure that they wanted to have pregnancies and be the mothers of little babies when they grew up, whereas one third of the fetally androgenized girls with the adrenogenital syndrome said they would prefer not to have children. The remainder, as well as the ten girls with a history of fetal progestin, did not reject the idea of having children, but they were rather perfunctory and matter-of-fact in their anticipation of motherhood, and lacking the enthusiasm of the control girls.

> When queried about the priority of a nondomestic career versus marriage and being a housewife in the future, the majority of fetally androgenized girls subordinated marriage to career, or else wanted an occupational career other than housewife concurrent with being married, and regarded occupational and marital status as equally important. Among the control girls, the emphasis was in favor of marriage over non-marital career. For the majority of these girls, marriage was the most important goal of their future. [101-102]

I have already suggested what might have occurred in the girls' environment to make them more interested in a career than in marriage. (What were their parents' expectations of them? What effect did their own knowledge of their anomalous sexual status have on the directions of their interest?) At issue now is how the authors measured their data. On what criteria were the control girls "rated high in enthusiasm for little children?" How was it determined that "they adored them and took every opportunity to get in close contact with them?" How do we know that the diagnostic group was "lacking the enthusiasm of the control girls" at the prospect of motherhood?

These questions are never answered, and arise again in their discussion of androgen-insensitive children. The subjects in this group were all genetic males. But due to a biochemical inability to utilize the androgen it produced, the fetus failed to differentiate male genitals. They were all assigned and raised as females, undergoing, with the aid of administered hormones, normal female puberty (except for the absence of menstruation, because of the absence of female gonads). Money and Ehrhardt present the following data in an attempt to prove that their medical histories did not prevent the subjects from developing what the authors regard as "normal" female gender identity:

> With respect to marriage and maternalism, the girls and women with the androgen-insensitivity syndrome showed a high incidence of preference for being a wife with no outside job (80 percent); of enjoying homecraft (70 percent); of having dreams and fantasies of raising a family (100 percent); of having played primarily with dolls and other girls' toys (80 percent); of having a positive and genuine interest in infant care, even though they had to forfeit the care of the newborn (60 percent); and of high or average affectionateness, self-rated (80 percent). Two of the married women each had adopted two children, and they proved to be good mothers with a good sense of motherhood. [111]

How did these women demonstrate to Money and Ehrhardt their "preference for being a wife with no outside job?" (Had they looked for jobs and failed to find them? Had they never looked at all, having resigned themselves to the unlikelihood of finding them, since they lacked necessary credentials? Had they never wanted a job because they feared even more than the "normal" woman snide remarks about being the one who wore the pants in the family?) What precisely is "homecraft" (washing dishes? needlepoint?) and how do we measure their enjoyment of it? (Because they do it with a smile? Or just because they do it at all?) How do we determine that their interest in infant care is either positive or genuine? And what exactly is a "good mother" or a "good sense of motherhood," and how does one prove to be the first and have the second?

Money and Ehrhardt do not provide answers to these questions; they do not even ask them. They offer us only a self-proclaimed objective eyewitness presenting hearsay evidence. Where in this is any outside point of reference? At least Erikson had a briefcase full of photographs we could pore over while we argued endlessly about their origin and their meaning. But Money and Ehrhardt give us nothing more concrete than *their* impressions of their subjects' self-impressions. When they conclude their description of the androgen-insensitive females (i.e., the genetic boys who were turned into social girls) by saying "the majority (90 per cent) of androgen-insensitive women rated themselves as fully content with the female role" (112), all they have given us is two networks of prejudice talking with each other in a common language. What questions do we ask to elicit that answer (e.g., "Are you fully content with the female role?")? And how do we evaluate their answers (e.g., "Yes I am fully content with the female role.")? Money and Ehrhardt accept this statement as proof-positive for the existence of such contentment in the diagnostic group, and by extension in "normal" females as well. Any amateur psychiatrist would be more sophisticated in the use s/he made of such "data."

It is hard to imagine, in America 1972, a scientist gathering together a group of any ten women and receiving without skepticism the information that nine out of ten were "fully content with the female role." In a society noted for its discontent with almost everything, should we not wonder at 90 per cent of any sample rating themselves "fully content" with anything? And what precisely is the "female role?" Is it the role of females in an impoverished Appalachian village? The role of females in a hunter/gatherer tribe in Africa? Of eighty-year-old widows in nursing homes? Money and Ehrhardt do not ever say, though we can probably assume (from their obvious orientation) that they intend something approximating the stereotype, to borrow their own phrase, of "shoddy popular psychology and the news media" (10): a white middle-class suburban housewife with some leisure time, spending money, and 1.9 small children. Possibly, we can even assume that the same was intended by those ten women. But whatever their individual intentions might be, the authors disregard the fact that the information is outrageously culture-bound, as is almost all of the other "scientific data" presented.

I alluded in the beginning to the authors' use of the word "tomboy" to describe the deviance displayed by the fetally androgenized females from their "normal" counterparts. Here is their up-to-date definition for that irrepressible word:

1. The ratio of athletic to sedentary energy expenditure is weighted in favor of vigorous activity, especially outdoors. Tomboyish girls like to join with boys in outdoors sports, specifically ball games . . . . Tomboyish girls prefer the toys that boys usually play with . . . .
2. Self-adornment is spurned in favor of functionalism and utility in clothing, hairstyle, jewelry, and cosmetics. Tomboyish girls generally prefer slacks and shorts to frills and furbelows, though they do not have an aversion to dressing up for special occasions. Their cosmetic of choice is perfume.
3. Rehearsal of maternalism in childhood dollplay is negligible. Dolls are relegated to permanent storage. Later in childhood, there is no great enthusiasm for baby-sitting or any caretaker activities with small children . . . .
4. Romance and marriage are given a second place to achievement and career . . . . The tomboyish girl reaches the boyfriend stage in adolescence later than most of her compeers. Priority assigned to career is typically based on high achievement in school and on high IQ. . . . [10]

It is possible that if one were to study exhaustively the role of women throughout history, one would discover certain common experiences in their lives (which is a long way still from discovering the reasons for commonalities). But it is likely that the list would *not* include "cosmetics," "frills and furbelows" (instead of "slacks and shorts"), "babysitting," "boyfriends," "priority of marriage over career," lower IQ scores, and all other such arch-American (vintage 1972) variations on femaleness.

One cannot study a fetally androgenized woman's choice of cosmetics as one would study the mounting behavior of fetally androgenized rats—a fact one futilely hopes would be obvious to scientists by now. A twenty-year-old woman's choice of cosmetics has been so thoroughly influenced by verbal and non-verbal communications from her environment since that time twenty-one years earlier when she received excess androgen *in utero,* that the only remote connection between the two events that one could postulate would be: Her high androgen level *in utero* caused her, *in some completely undetermined way,* to develop after birth some degree of identification with the male half of the population, which she *chose* to express by rejecting some of her culture's traditional signs of femininity. Given the dubiousness of the hypothesis, it seems strange that one would choose it to explain her deviant behavior, in preference to what we *know* was a factor in her development—that she and her family were aware of her abnormal condition, which required extensive medical treatment. But this is precisely the choice that Money and Ehrhardt make:

> The most likely hypothesis to explain the various features of *tomboyism* in fetally *masculinized* genetic females is that their tomboyism is a sequel to a *masculinizing* effect on the fetal brain. This *masculinization* may apply specifically to pathways, most probably in the limbic system or paleocortex, that mediate dominance assertion . . . *Masculinization* of the fetal brain may also apply to the inhibition of pathways that should eventually subserve *maternal* behavior. More correctly, one might say partial inhibition of these pathways, for normal males are capable of *paternalism,* much of which is identical with *maternalism,* both being manifestations of parentalism or caretaking. [103] [emphasis mine]

Their conclusion brings us to the third role of language in the study of sex and gender: how the language we use controls what we communicate.

## How Language Shapes Our Descriptions of Reality

The authors offer us above an evaluation of the extent of sexually differentiated behavior through a language itself thoroughly differentiated according to sex. Having accepted in the first conception (and title) of the book the dualities of man & woman, boy & girl, in which our language has codified our belief in a sex-dichotomous world, they then proceed to "prove" the existence of that dichotomy by fortifying it with all the cognates and derivatives of those two words that our language has spawned: male-female, masculine-feminine, masculinization-femininization, maternal-paternal. Is it not clear that their "most likely hypothesis" is nothing more than a tautology? In the act of naming the disease "tomboyism," they already impute to it a cause—namely, a woman imitating a man's behavior. They then name the possessors of the disease "fetally *masculinized* genetic females," thus doubly reinforcing the cause—of course, the "tomboy" is a "masculinized" female. Finally, they formally name the cause "a *masculinizing* effect on the fetal brain." Proof is executed: A equals A equals A.

They then digress briefly to a discussion of possible biological mechanisms involved in the presumed "masculinization," and return once more to the tried and truistic circular proof with the statement "Masculinization of the fetal brain may apply also to the inhibition of pathways that should eventually subserve maternal behavior. More correctly, one might say partial inhibition of these pathways, for *normal males* are capable of *paternalism,* much of which is identical with *maternalism,* both being manifestations of parentalism or caretaking" (emphasis mine).[11] We are thus presented with the peculiar task of measuring the extent of paternalism in fathers and maternalism in mothers to determine the degree of sexual differentiation in an activity, "parentalism," which the authors simultaneously admit *not* to be differentiated by sex.

The same fallacy underlies their statement elsewhere in the book that in the selection of playtoys, "boys tended to be more uniformly masculine in their preferences than girls were feminine" (181). And it underlies their entire discussion of dimorphic erotic behavior, in statements such as "Another extreme [in love patterns] is that of the Don Juan (in females the nymphomaniac) psychopathic type" (189). A neutral word could have been chosen to describe the same types of behavior in men and women, but once more it was not. So here, as in all the other examples, by choosing to assign different names to a characteristic when it occurs in men or in women, the authors guarantee that they will discover that characteristic to be differentiated by sex.

Money and Ehrhardt are careful to point out the fallacy of dichotomizing gender identity and gender role, as in the definition of "one type of male homosexual as 'having a masculine identity juxtaposed against a feminine role, that is a man who gives a masculine impression except that he relates erotically to a male, not a female'" (146). But they embrace the deeper fallacy of dichotomizing between masculine and feminine, as can be seen in their "correction" of the first mistake:

> Actually, such a person has an identity/role that is partially masculine, partially feminine. The issue is one of proportion: more masculine than feminine. Masculinity of identity manifests itself in his vocational and domestic role. Femininity of identity appears in his role as an erotic partner; it may be great or slight in degree, and it is present regardless of whether, like a woman, he receives a man's penis or, also like a woman, he has a man giving him an orgasm. It goes without saying that the ratio of masculinity to femininity varies among individuals .... [146]

*No*—it only goes with saying, and saying, and saying that there exist such polar concepts as masculine and feminine which, placed in varying ratios to each other, form the bedrock of our personality. It only goes with saying that a man who chooses to have sex with another man has a feminine identity as an erotic partner. We could just as easily name it spiritual, or bestial, or classically Greek, or anything else we like, and thereby alter our interpretations of the experience to conform to our preconception of it. But the masculine-feminine duality in behavior is the reigning construct of Western civilization. So, finding it difficult to fit male homosexuality into their conception of the "masculine" role, Money and Ehrhardt must therefore assign it to the "feminine" role— a strange solution, one would think.

The fundamental problem with accepting *a priori* the sexual duality as the primary construct of reality is that all of our discussions about sex and gender must then take place *within* this construct. All that the recent work on the relationship of sex to gender has done is move us from a consideration of human activity in two categories (male and female) to one in two pairs of categories (male, masculine; female, feminine). And we are left to spend our time squabbling over whether each trait displayed by a man is more rightly attributed to his maleness (sex) or his masculinity (gender).

Money and Ehrhardt make it clear that they accept the duality of gender identity as innate and indispensable when they draw their analogy to bilingualism. A child exposed to two different languages from birth is believed to assimilate both more quickly if they are spoken in two non-overlapping environments (e.g., one by father, the other by mother; or one at home, the other at school). In the same manner, the authors believe that a child will most easily acquire a healthy gender identity when exposed to two clearly demarcated gender roles in the parents:

> The traditional content of the masculine and feminine roles is
> of less importance than the clarity and lack of ambiguity with
> which the tradition is transmitted to a child. . . . When the models
> of gender identification and complementation have unambiguous
> boundaries, then a child is able to assimilate both schemas,
> the same way that a bilingual child assimilates two languages,
> the users of which are clearly demarcated and non-overlapping.
> The analogy with bilingualism is closer than it seems, if one consid-
> ers those cases of the children of immigrants who learn to listen and
> to talk in the language of the new country, but only to listen in the
> language of the old country. The parental language is enveloped
> with shame and indignity. . . . The two gender schemas are, in
> the development of the ordinary child, similarly coded as positive
> and negative in the brain. The positive one is cleared for everyday
> use. The negative one is a template of what not to do and say,
> and also of what to expect from members of the opposite
> sex. [164-5]

The ideal is for a child to have parents who consistently reciprocate one another in their dealings with that child. Then a five-year-old daughter is able to go through the stage of rehearsing flirtatious coquetry with her father, while the mother appropriately gives reciprocal directives as to where the limits of rivalry lie; conversely for boys. [186]

The authors' sense of the ideal and the appropriate in child-rearing leaves much to be desired. But while we may repudiate with ease the particular partisan interpretation which they have given to the sexual duality, we must still consider the existence of that duality in some form. So the question remains, is society's proclivity to view the world in a dualistic way supported by biological fact?[12]

There is good reason to suggest that the division of humanity into two *distinct* groups is at least in part a distortion of our biology. The widely believed conclusion Money and Ehrhardt reach through their comparative studies of hermaphrodites is that some sexually dimorphic behavioral traits have a genetic origin, most likely hormonally controlled. As has already been discussed, they never clearly state which particular traits they consider to be differentiated by sex (though they seem to fall into the traditionally accepted duality of male equals active and female equals passive). They do not ever attempt to verify the universal occurrence of these differences, with the exception of a brief and inconclusive study of erotic behavior in five primitive societies (see footnote 9). And they do not ever explain the mechanism by which they presume hormones to influence behavior in a predictable, universal way.[13] But even supposing all of these problems solved, sex and gender researchers approach the study of "sex" hormones with one more fallacy unexamined, as is manifested in the fact that estrogen is commonly referred to as the "female" hormone, and androgen as the "male" hormone. It is general knowledge that both estrogen and androgen are present in all males and females. It is perhaps not generally known how close, on average, the production of the two is in both sexes throughout most of our lives.

The only uniformly wide divergence is the elevated estrogen level in women during ovulation. The average level of androgen is somewhat higher in males than in females; the average level of estrogen (excluding the dramatic increase at the time of ovulation) is slightly higher in females than in males, until the age of fifty-five.[14] But average differences are rather small if we consider the ratio of the two numbers, and what effect those relative differences might have on individual behavior is purely speculative. The relationship of hormones to behavior is not well understood; and further, one would expect their impact on an individual to depend not on the absolute amount of hormones present, but on their relationship to the size of the individual and to the internal chemical environment of which they are only a part. Furthermore, within those average differences, there are wide variations that make it *not* uncommon to find women with higher androgen levels than those of the "average" man, and men with higher estrogen levels than those of the "average" (non-ovulating) woman. And finally, perhaps most importantly, androgen and estrogen are interconvertible in our normal body processes (see footnote 6).

And yet even Ann Oakley, in an otherwise excellent book, persists in calling androgen the "male," and estrogen the "female," hormone, leading to such verbal gymnastics as:

> [After age ten,] *in both sexes,* the production of *both male and female* hormones increases. With the approach of puberty, the increase in the production of *male* hormones *in both sexes* becomes pronounced. . . . The increase in the *female* hormone is much greater for *girls* than *boys* . . . but estrogen production in boys does increase at puberty. . . . While *men and women* produce *both male and female* hormones, the relative amount and proportions vary a great deal between individuals and one cannot establish biological maleness or femaleness from the hormone count alone. [24-26] [emphasis mine]

Wouldn't it make more sense, in studying the effects of "sex" hormones on behavior and physical development, to adopt the suggestion of Roger Williams that we classify humans into nine categories, representing all the possible combinations of low, medium, or high androgen and estrogen levels? There is far more at stake than semantics, because in recognizing the inaccuracy of the old terminology, we would recognize as well the inaccuracy of all of the sex-based generalizations that have followed from it. Even Williams' suggestion would not be precise terminology, because in talking about androgen and estrogen levels, we are discussing a phenomenon which is not dimorphic, tetramorphic or nonamorphic. It is di-spectral with wide overlaps, and so the number of points on each spectrum into which we choose to group humanity to study the effects of each hormone is arbitrary. But the larger the number of points we consider, the less distortion will be involved in fitting each person into one of the groups. And if we at least begin by dislodging androgen and estrogen from their inaccurate one-to-one correspondence to men and women respectively, we will have gone a long way toward clearing the air for honest scientific inquiry into their actual effects on behavior.

In the same way, the common division of somatotypes (body physiques) into female and male, paralleling the popular image of the "tall dark and handsome man" and the "lithe little woman," distorts reality. Most "secondary sex characteristics" are in fact *not* distributed dimorphically according to sex, but once again spectrally and with wide overlaps between women and men. Height, weight, musculature, body hair, breast size, pitch of voice are all characteristics that vary widely within each sex. While we find more men than women clustered at the tall end of the height spectrum, many women are there as well. And while the averages for the male and female populations diverge in all of these characteristics, it is important to realize that society is likely to have contributed even to this divergence 1) by establishing, over time, dimorphic criteria for sexual desirability; 2) by encouraging different diets and activities for women and men; 3) by encouraging people to mask their "deviant" characteristics (we can, for instance, guess that a large number of women have facial hair from the number of people who make their livings removing it); and no doubt by other mechanisms as well.

If in the end we limit our discussion of genetically prescribed dimorphic behavior, as Money and Ehrhardt finally do, to saying that "women menstruate, gestate and lactate and men don't" (*MWBG,* 163), we are at least approaching a truth (providing we amend that statement to read that *most* women menstruate, and *can* gestate and lactate for part of their adult lives if they so choose). What significance this difference has depends upon how much of her adult life a woman in fact spends gestating and lactating, and how much mental and physical effort is invested in each. And the more those decisions come to be personal choices for *each* woman, the less appropriate it is to generalize about their significance for *all* women.

Societally prescribed behavior, however, is quite another matter. Gender is by definition a dualistic concept, since the word is merely the symbol for our belief in a dualistic world. That belief has played a role so fundamental in ordering human experience that, in general, the lives of women and men in this and every other known society are undeniably different. Any attempt to study those lives must therefore be dual as well (with the proviso that if we analyze gender in its *public* terms of femininity and masculinity, we will never probe beneath its *public* image). But it is critical that discussions of sex and of gender be kept scrupulously unconfused.

Money and Ehrhardt's work is a self-proclaimed effort to achieve that clarity by "scientifically" determining the parameters of each. They have for the most part succeeded in confounding the truth even further. Indeed it may be an impossible task at this time. We know that cultural forces fuelling a sexually separate and unequal society are strong; one must question why scientists are so eager to ferret out whatever weak (and variable) genetic factors may be feeding the fire as well. Sex and gender researchers' predisposition to try, at a time when American women are once again coming to recognize the extraordinary impact of society on our destiny, seems a suspect effort to repin that destiny once more to our biology.[15]

## Notes

1. I am using the definitions of "sex" and "gender" used by Money and Ehrhardt in their book, *Man & Woman, Boy & Girl* (Baltimore: Johns Hopkins University Press, 1972). Subsequent references will appear in the text with the abbreviation, *MWBG*. "Sex" should be understood to refer only to the information transmitted to the individual genetically through the sex chromosomes. "Gender identity" is defined as "the sameness, unity, and persistence of one's individuality as male, female, or ambivalent, in greater or lesser degree, especially as it is experienced in self-awareness and behavior; gender identity is the private experience of gender role, and gender role is the public expression of gender identity" (p. 4). I think these definitions are themselves a distortion of the phenomena they purport to describe, for reasons that I hope will become clear in the course of this article.

2. For example, they attempt to isolate the various genetic causes of sexually dimorphic behavior by studying spontaneously occurring examples of human hermaphroditism. The groups they are studying are hence very small, ranging from a low of one, to a high of twenty-three members (pp. 98, 104, 106). Much of their evidence must therefore be considered inconclusive. Furthermore, it is questionable what a study of psychosexual differentiation under extremely stressful abnormal conditions tells us about the course of normal psychosexual development.

3. While Money and Ehrhardt in particular pay lip service to modern genetic theory that frowns upon the "antiquated dichotomy" of genetics and environment (p. 1), in fact, the greater part of the book is devoted to determining the definitive line dividing the two.

4. *See* for example Margaret Mead's classic, *Male and Female* (New York: Dell Publishing Co., 1968).

5. The problems inherent in cross-cultural studies will be discussed later in the paper. As for infants, they are a difficult population to study,

since they are relatively immobile, with a limited range of activities in which to display differentiation of any sort, sexual or otherwise. The statistics collected are further limited in their usefulness by the fact that they have already been affected by sexually dimorphic sensory stimuli administered by parents and others. For an excellent discussion of the methodological problems in infant studies, *see* Hugh Fairweather, "Sex Differences in Cognition," *Cognition,* 4 (1976), pp 231–80.

6. It is important to understand that both females and males secrete what we call "male" and "female" hormones—"androgens" and "estrogens" —which are regularly interconverted by our normal body processes. For an excellent discussion of the mechanisms of interconversion, *see* Ruth Bleier, "Myths of the Biological Inferiority of Women: An Exploration of the Sociology of Biological Research," *University of Michigan Papers in Women's Studies,* 2, No. 2 (1976), pp. 39–63. Bleier draws her data on hormone interconversion from two studies: K. H. Ryan et al., "Estrogen Formation in the Brain," *American Journal of Obstetrics and Gynecology,* 114 (1972), pp. 454–60; and Judith Weisz and Carol Gibbs, "Conversion of Testosterone and Androstenedione to Estrogens *in vitro* by the Brain of Female Rats," *Endocrinology,* 94 (1973), pp. 616–20.

7. For a spoof of this view, see R. Herschberger's *Adam's Rib* (New York: Harper & Row, 1970), especially chapter 8, entitled "Society Writes Biology."

8. Here, as elsewhere in the book, Money and Ehrhardt do not make explicit the intended scope of their thesis, though we are encouraged to view it as universal, through their implication that the difference is grounded in the mechanics of sexual intercourse. But whether their subject is America or the world, the only "proof" the authors offer for their assertion is, "There is good empirical support for this imagistic difference between the sexes, even in the absence of experimental design. The reader can test this hypothesis on the basis of his or her own experience" (p. 252).

9. This particular line of attack is useful in responding to some of the cross-cultural data Money and Ehrhardt present on "gender dimorphic traditions in sexual partnerships." They describe five different patterns of sexual partnership in primitive societies in order to show that although there is great variation between societies in the particular patterns prescribed, the patterns are invariably dimorphic (p. 145). Yet by their own indirect admission, their information is grossly incomplete. The researchers neglected to bring along a female investigator as part of their team, and since "talk about sexual activity . . . is taboo between the sexes" (p. 130), they obtained no information directly from the female members of each society. But rather than acknowledging the doubt this oversight must cast on the findings,

Money and Ehrhardt treat that lack of information as information: accompanying their lengthy discussions of prescribed and ritualized male homosexuality in these societies, they announce that "there is no evidence of a female homosexual relationship" (p. 133), and this in spite of the fact that they have admitted that such evidence would *not* have been available to the all-male team sent to collect it. One is of course forced to wonder if they would have been so ready to accept as definitive information any comparable *lack* of information about the male half of the population.

10. Erik Erikson, "Inner and Outer Space: Reflections on Womanhood," *The Woman In America*, R. J. Lifton, ed., (Boston: Houghton Mifflin, 1965), pp. 1–26. Ann Oakley, *Sex, Gender and Society*, (London: Harper & Row, 1972), pp. 96–97, 83. For another excellent critique *see* Kate Millett, *Sexual Politics* (New York: Doubleday & Co., 1969), pp. 210–220.

11. It is important to stress that all of this is entirely speculative. No neural pathways are known to exist that *do* subserve maternal behavior, much less "pathways that should eventually subserve" such behavior.

12. It has been a major contention of feminist theory (see, for example, de Beauvoir's *Second Sex*, Millet's *Sexual Politics*, Daly's *Beyond God the Father*) that men would find it in their self-interest to maintain this dualistic world-view with or without a biological basis for it. I think this is true, and serves in part to explain why John Money is predisposed to believe the duality of gender to be both innate and commendable, before he has proved it to be either. But that is a topic too large to be pursued further here.

13. This is a particularly distressing aspect of the book—that its argument is put forth in terms obvious enough that it is absorbed, but indirect and understated enough that one is not encouraged to confront it. No doubt, if pushed to defend some of the more spurious assumptions behind their argument, the authors would retreat safely behind the protection of having said the connection is only "presumed." But the fact remains that readers are allowed, indeed encouraged, to accept the connection as proved. And as the book progresses, through a predictable "carelessness," the presumed connection ceases to be treated as hypothesis, and comes to be regarded as fact. Thus, when discussing the adrenogenital syndrome, in which excess androgen is introduced into the female fetus, the authors say, "masculinization of the body is typically a source of mortification to the woman afflicted. The hormonal source of body masculinization does not also masculinize her mind" (p. 213). While ostensibly disproving one suspect fact, Money and Ehrhardt plant a far more outrageous one as fact in everyone's mind: i.e., "We still don't know exactly what masculinization of the brain is, but we now know it exists, though it is, by the way, not triggered by the adrenogenital syndrome."

14. Oakley, *Sex, Gender and Society,* pp. 24, 27.
15. In this light, it is not surprising that his research has led John Money
    to become one of the most outspoken proponents of transsexual
    operations, surely one of society's most reactionary institutional
    responses to the problems created by its sexual split. Considering
    the way in which doctors have dealt with homosexuality, some people
    have been surprised at the ease with which the medical establishment
    has granted transsexuality the status of normality. It should not be
    surprising at all: the establishment may have conceded a battle or
    two along the way, but it has won another war. Under the guise of
    the "revolutionary" insight that it is possible medically to cross the
    sex barrier, the proponents of transsexual surgery relink that barrier
    in the strongest possible terms to our biology. They affirm with all
    the sanctity of their profession behind them that there exists an
    *innate* difference between the *mental* lives of women and men that
    is so clear that it makes sense to speak of a "woman trapped inside
    a man's body." If there is any doubt about the reactionary implications
    of that belief, one need only look at the accounts male-to-female
    transsexuals have given of their own mental lives, many of which
    are grotesque parodies of our most degrading stereotypes of femininity.
    (*See* Janice Raymond's article, "Transsexualism: The Ultimate
    Homage to Sex-Role Power," *Chrysalis,* 3 [1977].) While transsexual
    operations may offer some sort of relief to a privileged few suffering
    an acute case of this society's sexual madness, they offer no comfort
    to those millions who find they are women in the right body,
    "trapped inside" the wrong society.

[Barbara Fried *is a doctoral candidate in English Literature at Harvard
University. In addition to being a freelance writer and editor, she teaches
writing in the Simmons College Graduate Program in Management. Her
published works include a critical study of William Faulkner entitled* The
Spider in the Cup *(Harvard University Press). She has been active in the
women's movement for many years, including as a manager of* Bread &
Roses, *Cambridge's first Women's Restaurant; and producer of the 1977–78*
Women of Words *series and other cultural events designed to bring women
artists and scientists to a larger feminist audience.*]

*Figure VIII. Katherine Dexter McCormick (1875–1967), early woman graduate in biology from M.I.T., at a suffrage parade in Chicago (1917). (M.I.T. Historical Collections)*

# Susan Leigh Star

## The Politics of Right and Left: Sex Differences in Hemispheric Brain Asymmetry

In vertebrates, the brain is divided into two halves—right and left—which in humans are thought to be specialized for different functions. Broadly speaking, some verbal and mathematical abilities are mostly controlled/processed through the left side of the brain, and musical ability, intuition, and spatial/Gestalt perception through the right side of the brain. This specialization is called *hemispheric asymmetry*, or *cerebral lateralization*. However, it should be noted that the left side of the brain receives messages predominantly from the *right* half of the body, whose movements it controls, and the right side of the brain monitors the sensations and controls the movements of the *left* half of the body.

There is a growing body of research asserting sex differences in the extent to which the left and right hemispheres contribute to brain functioning. Given the technical nature of research, "translation" and popularization to make it understandable are needed and have already begun. But both the academic research and its popularizations present dangers of concern to women.

Popularizations of current research on brain asymmetry rely heavily upon cultural stereotypes about left and right, mixed with Eastern and Western myths about yin/yang and left/right. Robert Ornstein, a major figure in both the research on brain asymmetry and its popularizations, draws upon the old Buddhist literature about yin and yang and left/right to emphasize his points about left brain and right brain functions. In so doing, he condones and reifies many traditional stereotypes of "masculine" and "feminine" and extends their use into "scientific" research.[1] Unfortunately, he is not unique in doing so.

Both Hinduism and Buddhism have posited different functions for the right and left *sides* of the body. The left side is named dark, mysterious, sinister and feminine; the right side is seen as light, logical and masculine. The Buddhist version says that the two sides of the body contain different types of energy, yin (left) and yang (right).

Such mythologies have found their way to the West as well. For example, the famous Osgood Semantic Differential Test, given in the late 1950's to a group of college students, showed the following common associations with left and right:

> The Left was characterized as bad, dark, profane, female, unclean, night, west, curved, limp, homosexual, weak, mysterious, low, ugly, black, incorrect, and death, while the Right meant just the opposite—good, light, sacred, male, clean, day, east, straight, erect, heterosexual, strong, commonplace, high, beautiful, white, correct. . . .[2]

Contemporary research suggests an association of left brain activity with verbal ability, and rational, linear thinking; and right brain activity with emotions, musical ability, Gestalt (holistic) ways of thinking, as well as with certain kinds of spatial ability. It is important to stress that phrases such as "left brain activity" do not necessarily imply exclusiveness, but may merely indicate an asymmetry in the activities of the two hemispheres that can be monitored electrically by recording brain waves, also called the electroencephalogram or EEG. Thus the left side of the brain has been identified through EEG studies as more active during speech. So scientists, like the myth-makers who came before them, have associated *sides* with *functions,* a phenomenon they refer to as lateralization.

*However, linking "feminine" and "masculine" to these functions is a false and stereotyped addition.* Often it is based on faulty methodology, sometimes on guesswork, both readily obscured by the technical language in which research findings are recorded. But unfortunately the (il)logic of patriarchy generates a dualism that equates women with the dark, mysterious, evil side of things, or else with the weaker, more emotional, less logical and less masterful side.

## Sex Differences Research

Before beginning a detailed analysis of the research and issues surrounding differences in brain asymmetry, several important issues must be raised about all sex-differences research.

1) In any research on sex differences, one must consider the possibility that potentials have been squelched in women by cultural stereotyping. Where biology is invoked to support such stereotypes, that "biology" should be critically and politically examined, particularly its language and research methods. In this area, it seems useful for feminists to unravel old myths and cultural stereotypes that have no factual

basis by using counterexamples and exposing the political and other biases of the research.

2) A more fundamental theoretical problem with sex differences research is that its basic orientation is toward *differences*. By searching these out it magnifies them and often obscures the fact that they may be the conveniently stereotyped extremes of broadly overlapping potentialities and functions. Research on sex differences therefore has often served to further stereotype and oppress women. The unstated biases that consistently inform this research are self-fulfilling: researchers find what they expect, and interpret their findings in traditional ways. Pseudo-scientific "proof" marching under the banner of "science" only strengthens old stereotypes.

3) One of the key ways in which this stereotyping has been perpetuated has been to confuse nature with nurture: anything that can be found to have a biological *correlate* is interpreted as innate. Patriarchal science conveniently equates biological sex differences with innateness, and innateness with genetic pre-determination (or tendencies/capacities with nature/function). Where biological correlates of sex differences are found, we must ask if they are a reflection of socialization—and if they could be a useful way to *examine* socialization (and implicitly, to counteract negative socialization). For example, if women's biceps muscles are less developed than men's, to what extent is this due to intrinsic biological differences and to what extent to differences in patterns of physical exercise of girls and boys from earliest infancy? One way of raising this issue is to ask: do we *somatize* our oppression? Rather than assuming that our bodies determine our social state, we must also consider how our social state shapes facets of our physical being, making both therefore changeable.

## Brain Asymmetry

What then should we accept as true about right and left brain differences? A specialization of functions seems to exist in each hemisphere of the brain. It follows therefore that if one side of the brain is damaged, the other side cannot always fill in, or take over. A common example of this is a bloodclot in the brain, a "stroke": if a stroke victim loses control over the movements of one side of the body, or over speech or other functions, the uninjured part of the brain does not take over immediately (although relearning is possible in varying degrees, depending on the extent and location of the damage).

Lateralization (asymmetry of brain function) is not as rigid in young children as in adults, though it appears to be fairly complete at five years of age—which is relatively late in the course of human brain development. Before this age, damage to either hemisphere often results in the other assuming the functions of the injured one. After five, major damage to one hemisphere can lead to permanent losses of ability.[3]

Initial studies on hemispheric asymmetry began with the work of Roger Sperry in the early 1960's. He and his co-workers studied surgical patients with severe epilepsy who had had their *corpora callosa* (the nerves and other tissue connecting the two halves of the brain) severed in the (successful) attempt to control seizures. As a result, these people had two separately functioning brain systems—their right hand literally did not know what the left was doing. By presenting the two sides of the brain with a variety of stimuli, Sperry was able to generalize about the types of functions that the two hemispheres perform separately. He concluded that the left brain determines logical thought, most speech, mathematical ability and "executive" decisions while the right brain rules spatial ability, emotions and intuitions. But even at this level, there was stereotyping: the left brain was called the "major lobe" in the literature[4]—despite the fact that the two lobes are of equal size and the same functional differences were observed in both sexes. The right brain was called minor and was perceived to be a "passive, silent passenger who leaves the driving of behavior mainly to the left hemisphere."[5]

Although Sperry found a clear-cut dichotomy between left and right brain functions, he and later investigators recognized that in normal, uninjured individuals the two hemispheres were interdependent and that right brain functions were no less significant or complex than those of the left brain. However the strict equation of spatial ability with right hemisphere functions and of verbal ability with the left, combined with the tenet that it is most efficient to use one side at a time, formed the basis for most subsequent theories about sex differences in asymmetry. Thus did theory become "fact" that was subsequently used to build new theories.

*Despite the demonstrated interplay between the left and right halves of the brain, verbal and spatial abilities had been so firmly identified with sides of the brain that many experimenters simply equated superior verbal ability with "left brain dominance" and superior spatial ability with "right brain dominance" without recording EEGs (brain waves) or performing any other physiological measurements to determine which half in fact was active during the specific tasks.* Therefore, most of the major hypotheses about sex differences in hemispheric asymmetry are inferred from differences in performance on specific verbal and spatial tasks and are based on little or no direct physiological measurements. At least four highly questionable assumptions underlie later experimenters' conclusions:

1) That certain tests (ranging from college board scores to various "spatial tasks") actually measure verbal or spatial ability;

2) That this measured verbal or spatial ability is the same as that which Sperry (and later, others) deduced from the surgically "split-brain," epileptic individuals;

3) That all spatial tasks "activate" the right brain and verbal tasks the left brain; and perhaps most importantly,

4) That differences in performance on tests are due to differences in intrinsic brain lateralization rather than to differences in prior training or experience, or differences in response to specific test situations.

Eleanor Maccoby and Carol Jacklin give an excellent summary of the problems with the first assumption in their book on sex differences.[6] There are serious questions about whether the verbal and spatial tasks/tests actually measure ability, as well as problems regarding sex differences in the taking and administering of these tests. For example, a female taking a "spatial" test from an older male researcher in a laboratory coat may well score lower than her male age peer for reasons other than intrinsic ability.

Assumption 2 is just that—an assumption. There is no way to be sure whether superior scores on college board vocabulary tests indicate greater left hemisphere dominance. Yet this assumption is rarely questioned in the literature.

Assumption 3 is especially tenuous when one looks at the variety of tasks that are categorized as "spatial." Everything from fine discrimination tests and detailed tasks to mechanical, manipulative tasks is grouped under "spatial" and is assumed to be associated with right brain activity—an association initially never tested with EEG measurements.

The final assumption, though rarely questioned, is clearly dubious in view of sex differences in training, especially for spatial and mechanical tasks.

The literature on brain asymmetry presents a classic case of sexist language, sexist interpretation of biological findings, and an interesting reflection of male-centered valuation. *Many of the observed sex differences can be easily attributed to socialization or faulty methodology . . . and these are often the sex differences that are interpreted in the literature as "male superiority"* (e.g. male "superiority" on spatial tasks for which boys traditionally have been trained).

The "Levy-Sperry Hypothesis" and the "Buffery-Gray Hypothesis" have been the two most widely discussed and believed theories about sex differences in brain asymmetry. They are both based on alleged sex differences in spatial and verbal tasks, and represent two entirely different reasonings from the same set of "facts": Levy and Sperry say that women are inferior on spatial tasks because of a *lesser degree of lateralization;* Buffery and Gray say that they are superior on verbal tasks because female brains are *more lateralized!*

## The Levy-Sperry Hypothesis

Levy and Sperry begin their reasoning by noting that females perform poorly on certain tests for spatial abilities, and that left-handed men perform poorly on the same tests. Left-handers, they state, perform poorly on these tests because of "cross-talk" from their left hemispheres while performing the tasks: they are said to be less lateralized.[7] The authors argue that the superiority of right-handed males in such spatial tasks is due to a *greater* lateralization of the two hemispheres; Levy states that "it might be that female brains are similar to those of left-handers in having less hemispheric specialization than male right-hander's brains."[8] She and Sperry also draw a further analogy between females and left-handers: they state that in left-handers language is mediated by both sides of the brain (whereas in right-handers it is a left brain function) and that the language component in the right hemisphere of left-handers (which is absent in right-handers) is what interferes with "pure" right-hemisphere performance on spatial tasks. (In fact, it is *not* true that left-handers usually have bilateral language representation.[9]) From this Levy and Sperry generalize to females who, they assume, also have bilateral representation for language, and they conclude that this is why females as a group perform more poorly on spatial tasks.

A number of researchers have already begun to accept their *hypothesis* as *fact,* and are using it to interpret further findings, although the problems with it are legion. Levy and Sperry do not address training and socialization as possible factors in performance of spatial tasks. They do not verify their assumption that the tests measure the degree of hemispheric specialization. And they do not address the critical fact that females consistently perform better than males on tests of *verbal* ability, a fact that would seem to contradict their assumption that females have bilateral language representation (which, by their reasoning, should make their verbal abilities *poorer*). Rather, they seem more interested in explaining male *superiority* on spatial tasks, whatever contortions of logic this might demand.

## The Buffery-Gray Hypothesis

Buffery and Gray examine the same test scores as Levy and Sperry, which show that males'perform better at certain spatial tasks; but unlike Levy and Sperry, they also take into account female verbal superiority.

To explain how both apparent superiorities could exist, Buffery and Gray construct the following hypothesis. They postulate that in males, linguistic and visuo-spatial abilities are represented in both hemispheres, whereas in females they are separated into the left and right hemispheres respectively. (Thus for Buffery and Gray, females are *more* lateralized than males, exactly opposite to Levy and Sperry's conclusion). Buffery and Gray then assert that bilateral representation is most efficient for visuo-spatial tasks—a direct contradiction of most theories—because they require a global, holistic perception. Hence males, with less lateralization than females, perform better on visuo-spatial tests. Then, with a confounding leap in logic, they assert that verbal tasks "require *more* lateralization," since they are more "specific" and "delicate" and "localized" than spatial tasks. Hence women, with greater lateralization than men, perform better at verbal tasks.

There are three serious problems with their hypothesis. The first is that attributing a more global or Gestalt perception to the spatial tasks that have been used in these cases requires a bit of imagination. One test used, for instance, is the rod and frame test—the ability in a dark room to adjust a movable glowing rod within a tilted frame to a vertical position. Another tests the ability to distinguish pictures of familiar objects within a camouflaging background. The ability to take a figure out of its background context is called field independence and is used as an example of "spatial ability."

In these sorts of tests females, on average, are less able to separate a figure from its immediate context, and are therefore said to be more field dependent than males.[10] From this it would appear that *females* are the ones who exhibit Gestalt perception (right brain), yet this is attributed to *men* in the attempt to explain their supposedly superior spatial ability.

But a more blatant contradiction emerges from the Buffery-Gray theorizing. They casually mention that male superiority on visual tasks only appears when manipulation of spatial relationships is involved. On tasks which depend for their execution principally on the discrimination and/or comparison of fine visual detail, the direction of the sex differences is reversed. Thus women are better than men on . . . a *number* of other tests of visual matching and visual search. . . . (emphasis mine)[11]

Thus, the *only* tasks that show men more able are tests of manipulation of the environment or of some part of it. The equation of this with spatial ability, not to mention its high valuing, reflects the respect accorded the skill of manipulation in this society.

Buffery and Gray end the above quote with: "Thus women are better than men on . . . a number of other tests of visual matching and visual search *which are predictive of good performance on clerical tasks*"![12] (Emphasis mine.)

Finally, Buffery and Gray, like Levy and Sperry, never verify the brain activity they associate with particular tasks. They *postulate* that men are less lateralized than women; they *postulate* that verbal skills require greater lateralization, and visuo-spatial skills less lateralization. But they never measure with brain wave recordings (EEGs) or other tests the actual brain activity of males or females during the performance of any of these tasks.

## EEG Studies on Sex Differences in Brain Asymmetry

In recent experiments Davidson and Schwartz have recorded the EEGs of men and women performing different tasks.[13] They find that right-handed females show greater brain asymmetry than males on self-generated tasks and can also control the *amount* of asymmetry more precisely. For example, when women are asked to perform a "right brain" task, such as whistling, they use more of their right brains and less of their left brains than do men. This is also true for left brain tasks. When asked (after prior training) to produce more or less asymmetry in either direction, females are much better at both tasks.

Another task was to produce an internal state of high emotion or a non-emotional state at different times. This was done by either re-living a situation of intense anger, or by thinking of some nondescript topic. During the emotional tasks, females showed more right-hemisphere activity than males; during the non-emotional tasks, they showed *less*. *Males showed no shift in asymmetry between the two emotional con-*

*ditions.* So, females showed greater shifts in asymmetry between emotional tasks and non-emotional tasks than did males in this study.

The third task involved biofeedback training to enable people to control the amount of hemispheric asymmetry they were producing and to produce asymmetry at will. Females were better at producing greater asymmetry, even to the degree of using *only* one side of their brains. There were no sex differences when people were asked to use both sides of the brain simultaneously.

Davidson et al. generate the following sets of hypotheses on the basis of their work:

1) They characterize the "male" way of *feeling* as much more analytical, more "left brain," while females may typically process emotions in a more global and Gestalt-like manner. In other words, during these experiments males were unable to produce a right-brain emotional state without left-brain interference.

2) When asked to think about something without emotion, males were less able to do so than females. This provides an interesting twist to the traditional stereotype that women's "emotional" way of thinking clouds their rationality.

3) Females have better control over the direction of their EEG asymmetry than do males—i.e. they can utilize *either* hemisphere more precisely depending on appropriateness.

Another piece of information directly relates to the nature/nurture question and brain asymmetry, although at this time its significance is not clear. In a study in which the brains of 100 newborn infants were measured at autopsy, Witelson and Pallie[14] reported that female infants have a significantly larger left temporal planum—the portion of the left brain that is primarily responsible for speech—than do male infants, in proportion to their total brain size.

## *Interpretations from a Feminist Perspective*

What seems most clear from the studies I have examined is that the imputation of male superiority on "visual-spatial" tasks, used as the basis for most of the theorizing about brain asymmetry, is a shaky and artifact-laden generalization. Most of the test results can probably be attributed to training or socialization, and do not necessarily reflect inborn differences in brain functioning. Furthermore, even if men's superiority were a proven fact, which it is not, the significance of the "visuo-spatial tasks" is questionable. What these researchers seem to be saying is that on tests of "spatial skill" men are better at manipulating the environment, except when "focused, delicate" spatial skills are required. Manipulation of the environment is not necessarily a desirable skill, nor is the ability to take things out of context (field independence).

For feminists, our central concern must be to eliminate patriarchal mechanisms that have blocked the expression *and validation* of language and spatial/environmental skills in women, and to encourage the development of those skills in the holistic manner of which we seem to be capable. There is no more poignant expression of the way in which women's capabilities have been distorted by socialization than the above quote about clerical abilities from Buffery and Gray. Similarly, the numerous adages about women's "excessive talkativeness" attest to the way in which our verbal abilities are often devalued.

New research paradigms are emerging in the areas of brain research and in other types of work in psychophysiology. Some questions that feminist researchers need to task are:

> What, if anything, do different ways of processing thought tell us about our potential and our socialization?

> How can we use our ability to feel deeply and think clearly in an integrated way?

> What kinds of spatial abilities do we want to develop and validate for ourselves? How can we strengthen the value of non-manipulative environmental skills?

> How can we enhance the optimal use of *all* our mental potentialities and skills?

One thing that emerges from the data is that men seem to have difficulty employing their right cerebral hemispheres in a focused way or for the execution of any but manipulative spatial tasks. Yet researchers have interpreted the test results to mean that men are superior in "spatial ability."

We are dealing with two contradictory stereotypes: men are linked with the right brain because of their supposed superior spatial abilities, women with the left brain because of superior verbal scores. Yet men are also stereotyped as "more analytic" (or "logical") which is said to be a left-brain skill. Complex and circuitous arguments are required to come up with "superior spatial ability" while leaving the myth of men's razor-sharp intellect intact.

Hemispheric asymmetry research at present is only a small part of the brain research that is being done on sex differences. But it is an extremely important part—both in itself and as an example of what can happen when psychological research is poorly designed and/or overinterpreted.

There is a strong relationship between attempts to dichotomize right brain/left brain ways of perceiving, thinking, feeling, and being in relation to the environment, and the dualistic social system summarized by the term patriarchy. In spite of the complex and often contradictory nature of the literature on left and right brain functioning, men's dominance in the social and economic spheres has been linked simplistically to their capacities for linear (left-brain) thinking.

This point of view is unfortunately adopted by some feminists in their condemnations of male dominance. For example, in her article in *Amazon Quarterly,* gina writes:

> So dualism resides in the very brain. The ways of perceiving
> that came to be grouped in the left hemisphere are the tools men
> used to take control of the planet. Linear thinking, focused
> narrowly enough to squeeze out human or emotional considerations,
> enabled men to kill . . . with free consciences. Propositional think-
> ing enables men to ignore the principles of morality inherent in
> all the earth's systems, and to set up instead their own version
> of right and wrong which they could believe as long as its logic
> was internally consistent. . . . All ways of perceiving that threat-
> ened the logical ways with other realities were grouped together
> on the other (right) side of the brain and labeled "bad."
> The separation of 'good' and 'bad' qualities into left and right
> sides of the brain, and the universally constant valuation of
> qualities, can be seen in every patriarchal culture through its
> attitudes toward left and right-handedness. . . ."[15]

Thus gina, in turn, introduces a dualism that *rejects* as male our ability to use the tools of intellectual reasoning and logic. This, too, is dangerous for it perpetuates stereotypic masculine/feminine dualities in an even subtler way by attributing them to the same person. Women's left brains are not precarious, "male" places to be visited but not dwelt in.

The dangers of greatest concern to women are that: a) researchers (men or women who unquestioningly accept their androcentric education) characterize women's and men's abilities on the basis of wrong research results and unfounded interpretations; b) these interpretations are already often uncritically accepted as true, and thus form the grid through which we then formulate further experiments and interpret their results; c) interpretations favorable to women are not drawn out of the data; d) some feminists are reacting to the popularizations of "left brain"/"right brain" thinking by misguidedly urging women to "reclaim our right brains," which shortchanges what women can do.

What is required is not the patching together of "left brain" with "right brain" qualities to form a pseudo-whole. Rather, the precise use of *all* our skills is what we should consider ideal.

## Notes

I would like to thank Artemis March and Richard Davidson for their help and advice.

The writing of this paper was partially supported by a Public Health Service training grant from the Human Development and Aging Program, University of California, San Francisco, No. AG00022-10 5T01

1. Note here that it is "feminine" and "masculine"—not female and male. This subtlety allows for a woman to be called masculine (or a male, feminine) if she exhibits behavior *a priori* classified as masculine. This reification eventually leads to the nonsensical concept of androgyny—two socially-created halves joined to make a "natural" whole.
2. Domhoff, W. "But why did they sit on the king's right in the first place?" *The Nature of Human Consciousness,* R. Ornstein, ed., (San Francisco: W. H. Freeman, 1973).
3. One of the many unknowns in the field is why the capacity to utilize either hemisphere for specific functions is lost after this age. In the research in this area, two intriguing findings are: a) the critical age for lateralization is later with lower socio-economic status, and b) females throughout life appear to retain more capacity for an undamaged right hemisphere to take over speech in the case of damage to the left hemisphere. For information on lateralization and socio-economic status *see* E. E. Maccoby and C. Jacklin, eds. *The Psychology of Sex Differences* (Stanford: Stanford University Press, 1974), p. 126. Information on females and lateralization was originally gotten from an article, gina, "Rosy Rightbrain's Exorcism/ Invocation," *Amazon Quarterly,* 2:4, 1974. Richard Davidson, a researcher in the area, agreed in conversation that it was correct. *See also*: H. Lansdell, *American Psychologist,* 16:448 (1961) and *Journal of Abnormal Psychology,* 81:255(1973). A thorough recent review of sex differences in brain asymmetry can be found in H. Fairweather, "Sex Differences in Cognition," *Cognition,* 4 (1976): pp. 231–280.
4. R. W. Sperry, "The Great Cerebral Commissure," *Scientific American,* January 1964, pp. 42–52; R. W. Sperry, "Split Brain Approach to Learning Problems," *The Neurosciences: A Study Program,* G. C.

Quarton et al., eds. (New York: Rockefeller, 1967); J. Levy-Agresti and R. W. Sperry, "Differential Perceptual Capacities in Major and Minor Hemispheres," *Proc. Nat. Acad. Sci., U. S.,* 61, p. 1151 (1968).

5. R. W. Sperry "Lateral Specialization in the Surgically Separated Hemispheres," *The Neurosciences; Third Study Program,* F. O. Schmitt and R. T. Wardon, eds. (Cambridge: MIT Press, 1974).

6. Maccoby and Jacklin, *The Psychology of Sex Differences,* pp. 125–127.

7. I have not yet found out whether EEG studies have been done on left-handers to determine if they *are* less lateralized during these tests or if this is another example of reasoning back from performance. It occurred to me while writing this that there could be a significant training factor involved with the performance of spatial tasks for left-handers, since many tools, games, and everyday spatial tasks are designed for right-handers. For example, my sister, who is left-handed, was poorer at performing the spatial task of cutting things out of paper with scissors—until she was given a special pair of left-hander's scissors, and my left-handed grandmother taught her to use them, after which she could perform the task with ease.

8. J. Levy, "Lateral Specialization of the Human Brain: Behavioral Manifestations and Possible Evolutionary Basis," *The Biology of Behavior,* J. A. Kiger, Jr., ed. (Corvallis: Oregon State University Press, 1972), p. 174.

9. J. Marshall, "Some Problems and Paradoxes Associated with Recent Accounts of Hemispheric Specialization," *Neuropsychologia,* 11 (1973), pp. 463–470.

10. Lauren Gibbs, "Sex Differences in the Brain: Questions of Scientific Research," mimeographed (Cambridge: Harvard University, Department of Psychology and Social Relations, 1976): " 'Dependence-independence' are highly charged words in our society with strong value and sex referents. Independence is a prized character trait and associated with "masculinity," while dependence, a "feminine" trait, is despised. The unqualified use of these terms encourages continued acceptance of destructive male-female stereotypes and goals (less rigid reliance on the isolated self and less blocking out of context might be useful to our society) . . ."

11. W. Buffery and J. Gray, "Sex Differences in the Development of Spatial and Linguistic Skills," in *Gender Differences: Their Ontogeny and Significance,* C. Ounsted and D. C. Taylor, eds. (Edinburgh: Churchill Livingstone, 1972) p. 127.

12. The use of the word "predictive" here is interesting—it is an adjective with a deleted agent. This means that unmasked questions are implicit in the word, such as *who* predicts, for *whom,* and *why?* For a linguistic analysis of "scientific" terminology like this, which deletes agent, subject, and motivation, *see* Julia Stanley, "Passive Motivation," *Foundations of Language,* 13 (1975), pp. 25–39.

13. R. J. Davidson and G. E. Schwartz, "Patterns of Cerebral Lateralization during Cardiac Biofeedback versus the Self-regulation of Emotion: Sex Differences," *Psychophysiology*, 13 (1976), pp. 62–68.
14. Witelson and Pallie, "Left Hemisphere Specialization for Language in the Newborn," *Brain*, 96 (1973), pp. 641–646.
15. gina, "Rosy Rightbrain . . ." *See also* "Codifications of Reality: Lineal and Nonlineal" by Dorothy Lee in Ornstein, ed., *The Nature of Human Consciousness* for a description of a somewhat alternate culture.

[Susan Leigh Star *is poetry editor for* Sinister Wisdom *and West Coast Editor of* Matrices: Lesbian Feminist Research Newsletter. *As a lesbian feminist poet and theorist, she has been active in the women's movement for several years. She is currently at the Human Development and Aging Program, University of California, San Francisco. Her research interests include the psychology of consciousness, aging and ageism, and moral/ethical development.*]

# Part Two:

## Gaining Control

*Figure IX. Investigators on collecting trip, Marine Biological Laboratory, Woods Hole, Massachusetts (1895). (Society for the Preservation of New England Antiquities; Baldwin Coolidge, Photograph)*

# Introduction

*"You all die at fifteen," said Diderot,*
*and turn part legend, part convention.*
*Still, eyes inaccurately dream*
*behind closed windows blankening with steam.*
*Deliciously, all that we might have been,*
*all that we were—fire, tears,*
*wit, taste, martyred ambition—*
*stirs like the memory of refused adultery*
*the drained and flagging bosom of our middle years.*
                    —Adrienne Rich, "Snapshots of a Daughter-in-Law"

We see ourselves as others see us. Our visions are guided by the available options. These homey truths hold for women and men, but with an important difference. Little boys are perceived from birth as potential butchers, bakers, and candlestick makers; as future astronauts, machinists, garbage collectors, doctors, firemen, lawyers, policemen, mailmen, or politicians. The fact that many will one day be husbands and fathers as well has little impact on how parents and teachers perceive, encourage, and train them.

For girls, the story is quite different. Even in today's more liberated times they are perceived first and foremost as future wives and mothers. Today's perceptions permit, in addition, a limited range of professions, most of them semi-nurturing: nurse, teacher, hostess, secretary. Each requires above all else the skill to adjust one's needs to those of others, to submerge, subdue, efface one's self. So little girls have been taught traditionally; so they are still taught—though perhaps now more through silence than words.[1]

It is no wonder that, as Darwin noted, so few women had attained "higher intellectual eminence." When one considers the pressures brought to bear against them, the wonder is rather that any attained eminence at all—that any of our foremothers survived the distractions, the threats, the warnings of inevitable failure and the vicious contempt

I'd like to be an astronaut
And live in a space station.

If I become a secretary
I'll type without mistakes,

I may even be a housewife
Some day when I am grown,

And I'd love to be a mother
With some children of my own.

*Figure X.  Illustrations from two children's books*: What Girls Can Be *and* What Boys Can Be.

for their fledgling successes, to see their visions through to life and public light. M. Carey Thomas, founder and first president of Bryn Mawr College, gave the following account of those pressures on her early life; her account differs from those most of her contemporaries could give only in its happier resolution:

> I was always wondering whether it could be really true, as everyone thought, that boys were cleverer than girls. Indeed, I cared so much that I never dared to ask any grown-up person the direct

question, not even my father or mother, because I feared to hear
the reply. I remember often praying about it, and begging God
that if it were true, that because I was a girl I could not successfully master Greek and go to college and understand things
to kill me at once, as I could not bear to live in such an unjust
world. . . . I can remember weeping over the account of Adam
and Eve because it seemed to me that the curse pronounced on
Eve might imperil girls' going to college; and to this day I can
never read many parts of the Pauline epistles without feeling
again the sinking of the heart with which I used to hurry over
the verses referring to women's keeping silence in the churches
and asking their husbands at home. . . . I can remember one
endless scorching summer's day when sitting in a hammock
under the trees with a French dictionary, blinded by tears more
burning than the July sun, I translated the most indecent book
I have ever read, Michelet's famous—were it not now forgotten, I
should be able to say infamous—book on woman, *La femme*. I
was so beside myself with terror lest it might prove true that
I myself was so vile and pathological a thing.[2]

We are already in danger of forgetting the tangible horrors of our past;
of mistaking the real tears Thomas wept over Eve's fate for metaphorical
ones, of thinking that the terror engendered by Michelet's book had
some other vaguer, deeper, less prosaic cause. These men and the books
they wrote were real, and the toll they took on women's lives, though
immeasurable, was also real.

We are also in danger of forgetting how much the past is repeated
in present, subtler forms—of forgetting how fragile and often two-edged
are our gains. This section begins with two articles that illustrate that
what we have already secured can be lost again. It is a somewhat different
story from the official version of our society's steady, good-hearted
progress over the centuries towards the full liberation of women. Datha
Brack's paper traces how our birthing experience, once the province
of the mother and midwife, was professionalized away from us. She
also analyzes how some of the "improvements" on childbirth recently
instituted by men like Lamaze and Leboyer lead us not back to ourselves, but to a more insidious—albeit a more "hip"—form of alienation.
Mary Roth Walsh's paper explores the inroads women were able to
make into the medical profession in the nineteenth century, and their
subsequent exclusion that allowed a male monopoly over health care
for another hundred years.

The three articles on anorexia, menopause, and menstruation describe the ways in which our self-images incorporate society's manipulations of our bodies until it becomes impossible for us to know our selves.

In the final essay, Weisstein gives her view of what it is like to be young, gifted and female in the scientific community. Many, especially among younger women, may think her story is an old and finished chapter in women's history—to which we would respond, "Watch out." There are many ways to neutralize (and neuterize) women. One is: to keep us out. Affirmative action may prevent that, though it is far from clear that it will. But even if it does, women can also be rendered harmless by granting us full membership in the club provided we accept its rules.

Others, reading Weisstein, may be tempted to conclude that if she had all that trouble, it must surely somehow be her own fault. To blame the victim is one of the prime messages of our socialization; it conveniently leaves blameless those who hold power and make the rules. No: the victims are set up to lose, not because they are born losers, but because others have the power to keep them from winning. In a society that institutes elaborate rituals of equal access, this can sometimes be hard to see; indeed one of the functions of the rituals is to obscure this fact. Yet each of us, whether we stand outside or have managed to fight our way in, must try to become conscious of the many "keep out" signs that decorate the entrance.

Enter we must, but as Weisstein stresses, we cannot afford to do so as isolated individuals who have "made it." If we do, our entry will make no difference and will be but a further exercise in denial for ourselves and for those who succeed us in our token successes—for all the Cinderellas who will "die young/ like those favored before [us], hand-picked each one/ for her joyful heart."[3]

*December, 1977*

## *Notes*

1. These, incidentally, are close to the roles which that champion of "natural man," Jean Jacques Rousseau (1712–1788), laid out for his exemplar, Emile, and for Emile's perfect helpmate and companion, Sophie; roles that provoked Mary Wollstonecraft in 1792 to publish in protest *A Vindication of the Rights of Woman.*
2. M. Carey Thomas, "Present Tendencies in Women's College and University Education" *The Educated Woman in America,* Barbara M. Cross, ed. (New York: Teachers College Press, 1965), pp. 158–160.
3. Olga Broumas, "Cinderella" *Beginning With O* (New Haven: Yale University Press, 1977).

*Figure XI. Pre-Columbian bowl from the Mimbres Valley in New Mexico.*
*(Peabody Museum, Harvard University)*

# Datha Clapper Brack

# Displaced–The Midwife by the Male Physician

From earliest times, women have assisted other women in childbirth. This was true for preliterate peoples, agricultural societies, and our own Western culture prior to the Industrial Revolution. Even today, considerably more babies throughout the world are delivered by midwives than by physicians. Then how did it happen that in our society the female midwife was replaced by the male obstetrician?

Before this could come about, two fundamental conditions had to be met. First, men had to be defined as socially acceptable persons to attend childbirth, and second, normal childbirth, traditionally regarded as a "natural" event, had to be redefined as one that called for professional medical assistance. However, although these conditions allowed men to enter and practice midwifery, they did not require them to dominate the occupation; on the European Continent and in Great Britain, medically trained midwives continued to practice, and today they deliver sixty to eighty per cent of the babies there.[1] But in the United States, by the twentieth century, the midwife had all but disappeared. During the 1960s, nearly ninety-nine per cent of American babies were delivered by physicians, more than ninety-three per cent of whom were male.[2] When a joint study group of the International Confederation of Midwives and the International Federation of Gynecology and Obstetrics compiled a report of maternity care throughout the world in 1966, the writers found it necessary to treat the United States as a "special case" in the tables "because of its tendency not to recognize midwifery as an independent profession."[3]

It is likely that professionalization of medicine allowed male physicians to break the midwife's traditional hold of the occupation. As medicine developed in Europe from a learned calling to a consulting profession, the prestige of the profession allowed the physicians to make claims to superior competence as childbirth attendants, and co-opt the developing medical technology and knowledge applicable to childbirth. In the process of developing their expertise, they redefined the nature of childbirth and reorganized childbirth care to suit their own professional needs.

However, it was *as a professional group* that physicians in the United States subsequently were able to protect their own interests and dominate the field. Powerful medical associations gave them access to the power of the state, through which they were able to control licensing legislation, restrict the midwife's sphere of activity, and impose legal sanctions against her. Male physicians blocked women not only from delivering babies as midwives, but also from delivering them as practicing physicians. The exclusionary practices of medical associations and the use of informal professional networks worked both to bar women from studying medicine and to reduce their effectiveness as practitioners when they were qualified.[4]

To a certain extent, these pressures existed on both sides of the Atlantic, but additional social and cultural factors combined in the United States to give physicians an advantage. First, there had been no established tradition for training midwives in the United States such as had developed in Europe alongside the profession of medicine. This European tradition gave professional midwives more access to developing medical knowledge and more legal protection than their American counterparts had. Second, medicine itself was a newly imported profession in America, one that was having difficulty staking its claims against those of lay practitioners of all kinds in the nineteenth century. Its practitioners were therefore more jealous of what they saw as their territory than were members of the older, established profession in Europe.[5] Third, women were more likely to be restricted from entering prestigious professions in the United States. And finally, perhaps most importantly, at several critical periods, physicians were in greater supply here than in Europe. It was during these periods that attempts were made to restrict delivery of babies to the medical profession. At times when physicians were in short supply, midwives were more likely to be tolerated and suggestions for establishing and improving schools for their training more likely to be entertained.

## The Social Organization of Childbirth Care

It is useful to think of a profession as an occupation which has assumed a dominant position in a division of labor, so that it gains control over the substance of its own work. Unlike most occupations, it is autonomous or self-directing. In developing its own "professional" approach, the profession changes the definition and shape of problems as experienced and interpreted by the layman. The layman's problem is recreated as it is managed—a new social reality is created.[6]

It was necessary for medical men to change the definition of childbirth in order to manage it, because it had been institutionalized as a social event rather than an illness, and one that specifically demanded female participation. During childbearing, a woman passes through a sequence of exclusively female social statuses—pregnant woman, laboring woman, nursing mother—and for each of these, she has to learn role behavior expected by her own culture. In preliterate societies, experienced older women were the holders of wisdom about childbearing. They provided role models for behavior, taught the new mother what she needed to know, and gave emotional and social support during the birth event.[7]

When midwifery became established as an occupation in traditional society, it is not surprising that it continued to be woman's business. Midwives generally came from the social class of the women they served and therefore shared the same cultural expectations for childbearing behavior. They provided counselling and personal support not only at the birth event, but also before and during labor, and through the post partum period. Among common people, as a rule, the midwife moved in with the family when labor was imminent, attended the birth, stayed afterward to nurse the mother and child, did light housekeeping until the mother was able, and gave advice on infant care and breastfeeding. Even today's highly professional midwife gives care that is characterized as "sympathetic," "personal," "counselling," "total."[8] A recent article on midwives in the *New York Times* was captioned, "Delivery by an Old Friend."

A professional man could hardly have served the same functions. As the physician's role developed, it ideally demanded emotional detachment and services limited to the specific condition being treated.[9] Consequently, as obstetrics became a specialty, labor and delivery were isolated from the rest of childbirth care and defined as medical and surgical events. Development of modern hospitals and transportation made it feasible for childbirth to be moved out of the home away from its previous family and social context, and placed in the physicians' domain where they had access to the increasingly complex technology of obstetrics. In the United States in 1968, 98.5% of all deliveries were performed in hospitals.[10]

Furthermore, childbirth is spontaneous and unscheduled. Any particular delivery has its own timetable for day, time, and length of labor.

This presented no problem for the midwife, who was prepared to settle down and wait. But physicians were able to perform efficiently only as the profession developed a technology that allowed them more control over time and place. A study by Rosengren and De Vault shows how anesthesia is used in the modern delivery room to hurry or slow deliveries. The authors add: "The use of forceps is also a means by which the tempo is maintained in the delivery room, and they are so often used that the procedure is regarded [by physicians] as normal."[11]

A third characteristic of childbirth is that most deliveries are uncomplicated.[12] Although percentages of complicated births can vary depending upon a number of social and cultural factors, approximately 95% of deliveries—even in modern society—require little or no technical assistance other than catching the baby and tying the umbilical cord, chores the mother herself may perform quite casually in less sophisticated cultures. (And as police manuals point out, they are chores any patrolman can handle.) The origins of names for childbirth attendant suggest that the role is mainly one of waiting and supplying support. "Midwife" is derived from the Old English "with woman;" "obstetrician" from the Latin "to stand by;" "accoucher" from the French "to put to bed." The midwife waited for childbirth to take its natural course, intervening only in cases of abnormality; the physician developed care that stressed intervention rather than waiting. Freidson has pointed out that it is "the task—indeed the mission—of the medical men to find illness," that the profession is "first of all prone to see illness and the need for treatment more than it is prone to see health and normality."[13] In the United States, the profession came to define *all* childbirths as not only potentially complicated, but also as in need of "aid"; as a condition that ideally should always be hospitalized, medicated, anesthetized, and require an episiotomy.[14]

Finally, childbirth is associated with a woman's sexuality. Pregnancy begins with coitus, it alters the balance of female hormones, and the infant may nurse at the mother's breast—an organ invested in Western culture with strong sexual meaning. Furthermore, a woman's genitals are exposed and touched during examination and delivery—acts that in any other social context would be invested with erotic meaning. Mead and Newton point out that sexually allied emotions may be aroused at the time of delivery in both the attendant and the parturient, and that although this is often recognized in less sophisticated societies, it is taboo in America.[15]

While the sexual connotation of childbirth posed no problems for women attending women, it presented a considerable barrier to men.

Tracts published against men entering midwifery often expressed outrage at the impropriety of having any man other than the woman's own husband present while she was giving birth. This attitude remained a prime obstacle to the immigrant woman's acceptance of the American doctor as late as the 1920s. The social organization of modern obstetrical care denies or hides the sexual implications of childbirth, and thus minimizes or eliminates role conflict for the male physician. The removal of the woman from a familiar environment; the emotional neutrality of the doctor-patient relationship; the rituals of aseptic technique and surgical preparation; the routine use of medication and forceps; and finally, the routine use of anesthesia, which removes the woman from participating in the birth act—all place a medical rather than a social or sexual connotation on childbirth. It is, perhaps, a measure of the "success" of the medical profession in de-sexualizing childbirth that the claims of some prepared childbirth advocates to sensual and orgasmic experiences during delivery can be viewed with amusement and disbelief, and easily dismissed.

## Men in Midwifery—Conflict and Controversy

By the 16th and 17th centuries in Europe, the midwife held a recognized position in society and was sometimes well-educated and well-paid. She acquired her knowledge by formal apprenticeship to other midwives, was licensed to practice by the bishop, and successful midwives were considered little, if at all, inferior to doctors. Early textbooks were written by and for women midwives, and were used in the early universities.[16]

When men entered midwifery in the 17th century, they did so as university-trained physicians, and practiced among upper-class women who were accustomed to being attended in illness by men from their own class. The early professions (law, medicine, theology) had developed in the early university; the university in turn had developed within the powerful church, the seat of learning during the 12th and 13th centuries. The university was supported and attended by members of the aristocracy, and the professions became as strongly male-dominated as the patriarchal elite classes and the church from which they developed. By the 17th century, when men first entered midwifery, medicine was an established male profession in Europe, and the power structure of the time legitimized the physician's claim to expertise.[17]

But still they encountered obstinate barriers, for social institutions, laws, and customs all legitimized the woman midwife's control over delivering babies. For example, physicians were required to obtain special licenses to practice midwifery in London, and as late as 1762, when the first instruction in midwifery was given at the University of Edinburgh Medical College, it was confined solely to female midwives.[18] A midwife, Mrs. Elizabeth Nihill, gives the following account of the opposition physicians met when they attempted to practice midwifery in a public hospital in Paris. The men had learned their skills from midwives:

> and yet have many of those very men-practitioners, influenced
> by that self-interest which has such power in all human affairs,
> revolted against their mistresses in the art, and their benefactresses.
> They have, at various times, commenced law suits about the
> Hotel-Dieu at Paris, in order to get lyings-in there committed
> to them: but the administrators, the persons of a just sense
> of things, together with the parliament of the town, ever attentive
> to decency . . . have constantly opposed and frustrated the
> pretensions of these innovators. These again thus disappointed,
> were forced to content themselves with practising upon some
> women of quality under the favor and protection of some of the
> old ladies of the court of Louis XIV who had their reasons for
> the propagation of this fashion.[19]

Then, Nihill explains, physicians began to "run down" the midwives while "exalting themselves," and the "novelty prevailed," eventually spreading to the provinces.

Among common people, men entered midwifery through the authority of the guild. Sarah Stone, an English midwife, states that "almost every young man who has served his apprenticeship to a Barber-Surgeon immediately sets up for a man-midwife; although as ignorant, and indeed much ignoranter, than the meanest woman of the Profession."[20]

Although the authority of the professions and the guilds made it possible for men to gain a foothold in midwifery, it was the obstetrical forceps, developed early in the 18th century, which encouraged them to enter the occupation in large numbers and challenge the midwife's dominance. Using forceps, a birth attendant could extract the infant during labor, gain control over the timing of delivery, and thus operate more efficiently.[21]

The ascendancy of men-midwives precipitated a controversy that raged in popular and medical literature for the better part of two centuries. It persisted until medicine had firmly established its monopoly over healing and its control over obstetrics.[22] The use of instruments—

principally obstetrical forceps—was a pivotal issue around which claims of superior competency and charges of abuses turned. The "natural right" of women to attend women in childbirth and the "immorality" of having a man attend delivery also figured prominently in the debates. And finally, the profit motive—a persistent theme in any story of occupational displacement—pervaded the controversy.

At first there were no clearly drawn battle lines between midwives and male practitioners. Midwives defended men-midwives when they were competent,[23] and there were male physicians who saw women as more capable in the occupation than men.[24] Forceps and other instruments were used by trained midwives as well as men practitioners. The difference was that midwives still waited for a delivery to take its normal course and used instruments only in extreme cases. The men were accused both by midwives and by other physicians of using them indiscriminately to hasten the delivery and make a profit, damaging mothers and infants in the process.[25] But other physicians defended the use of forceps as an advance, referring to them as "noble and beneficient" instruments that had "rescued many lives."[26] Early obstetricians were presented as brave men who were pioneering in improvements while they withstood the attacks of the midwife who saw her livelihood slipping away from her.

Pamphlets debating the morality of the issues bore such colorful titles as the following (both published mid-19th century):

> *Medical Morals. Designed to show the Pernicious Social and Moral Influence of the Present System of Medical Practice, and The Importance of Establishing Female Medical Colleges.*[27]

> *Man Midwifery Exposed; or The Danger and Immorality of Employing Men in Midwifery Proved; and the Remedy for the Evil Found.*[28]

And finally, the vaginal examination practiced by physicians came under attack both by medical men who spoke of the danger of introducing infection and by the pamphleteers for whom the sexual implications were frequently just as worrisome. For example:

> Where then is the excuse for the indecencies and outrages of man-midwifery? Why is it that, without plea of necessity our wives are exposed to the shame and pollution of examinations which are *invariable,* and manipulations such as a pure-minded and sensitive woman must blush to think of—such as must excite

the indignation of every man who regards the person of his wife as sacred?[29]

However, by early 20th century, as medical knowledge became more advanced, the controversy had shifted to other issues: to responsibility for the high maternal and infant death rates, and the problem of inadequate training both for midwives and for physicians.

## The Dominance of the Medical Profession over Childbirth

In the case of medicine a significant monopoly could not occur until a secure and practical technology of work developed.[30]

At the beginning of the 19th century, the claim of the medical profession to practice midwifery was considered legitimate regardless of the controversy surrounding the physician's management of delivery, but the midwife was still an independent practitioner. By the end of the century, medicine's preeminence among the healing occupations had been established, and its dominant position over the occupation of delivering babies was assured. While in England and the rest of Europe this dominance meant that the medical profession controlled most decisions about the licensing of midwives, and about which medical and surgical procedures licensed midwives could legitimately perform, in the United States it meant a slowly developing monopoly of the occupation.

Generally, in Europe it was assumed that childbirth would be normal and therefore the responsibility of trained midwives, unless there were indications that a delivery might be complicated. The midwife's role specifically included the obligation to recognize indications of abnormality, and turn such patients over to the physician. In actual practice she was trained to handle emergencies, and it was implicitly recognized that she would do so "on her own responsibility" if a doctor's assistance were not available on time. In the United States, on the other hand, all deliveries came to be seen as potentially complicated and treated as such. They were therefore the province of the physician. As the 20th century progressed, the tendency was to treat all deliveries as surgical events: i.e., ideally they were performed in a surgical operating room of a hospital, with the "patient" surgically prepared (enema, surgical shave, no food), and with surgical interventions (medication, anesthesia, routine episiotomy).

A number of factors contributed to medicine's domination of childbirth management. One was the increasing sophistication of the medical and surgical knowledge applied to the complications of childbirth, much of which was progressively restricted by law from use by mid-wives. Anesthesia, medications, laboratory techniques for predicting complications, all were made more available to physicians than to midwives. A second factor was the dramatic increase in large public hospitals, a development that reflected the widespread social change at that time. The Industrial Revolution brought increasing urbanization, with its concomitant poverty and crowded ghettos of the urban slums. Women as well as men worked long hours under brutal conditions and were poorly nourished. Sewage systems and sanitation were inadequate, and in the summer poor refrigeration made the contamination of food likely. Under these conditions, disease rates for the entire population were high, and maternal and infant death rates soared. The situation stimulated widespread concern over public health and resulted in the building of hospitals with both public and private funds. In the United States, in the fifty years between 1873 and 1923, the number of hospitals increased from 149 to 6,762.[31] It was in these hospitals that medicine developed its expertise in obstetrics.

Finally, during the 19th century, the profession of medicine organized powerful medical associations that grew to exert strong influence over directions of health care. These associations gave the

*Figure XII.   Women's Laboratory at M.I.T.*
*(M.I.T. Historical Collections)*

profession the power to dominate obstetrics as a medical specialty and to define its scope, and in the United States, the leverage to monopolize it.

## Occupational Displacement in the United States

Although it is likely that in this country midwives still delivered over 50 per cent of the babies early in the 20th century, they were an unorganized, unrepresented group of women who had little social prestige or power.[32] By and large midwives served lower-class immigrant women of their own ethnic groups whose cultural expectations led them to seek midwives; indigenous "granny" midwives served women in out-of-the-way places such as the Kentucky hills and the deep south. These patients would have been unprofitable for a private physician. While most state laws were requiring medical men to be licensed, and, at least by the second decade, medical associations were setting more rigid standards for training and licensing, there was little uniformity and control over qualifications to practice midwifery. Only six states required a midwife to be trained; only twelve plus Washington, D.C., required her to pass an examination. In seven, the regulations were inadequate, and in fourteen, there were no laws at all governing her training, registration and practice. Since she had no medical degree, laws in most states prohibited her from giving medication or performing even minor surgery. Furthermore, while medical school standards were growing more exacting, few quality schools for midwives had survived from the past century.[33]

By comparison, in Great Britain and elsewhere in Europe, midwifery was a profession that was still recognized and respected. In most cases, midwives were required by law to be trained and licensed, their practice was regulated, and as professionals, they were allowed to collect professional fees for services. Their services were provided to the poor at public expense. Excellent schools with sound medical training were available to them.

England's situation had been developing similarly to that in the United States, until the Midwives' Act was passed by parliament in 1902.[34] The act provided for the establishment of a Central Midwives' Board, required all midwives to be trained, and empowered the Board to regulate their training and supervision.

The question remains why the situation developed one way on one side of the Atlantic and another way in the United States. One factor may have been a difference in cultural history. Midwifery had at one time been the occupation of women of education and prestige in England, trained and licensed by the state, so that the profession had both tradition and official sanction to give it legitimacy. The reverse was the case in the U.S.

A second factor may have been a difference in cultural attitudes about women. Cynthia Epstein has pointed out that the Victorian "idealization" of women that depicted them as frail, in need of protection, and not very capable (middle-class, not working-class, women), had a greater hold in this country than elsewhere, and was one of the factors that prevented women from entering professional life.[35] This attitude pervades the literature of the midwife controversy throughout the 19th century, and was used as an argument for discouraging women from studying medicine as well as from practicing midwifery.[36] The following example from a pamphlet published in 1820 by a Boston physician is typical:

> I do not intend to imply any intellectual inferiority or incompetency in the sex. My objections are founded rather on the nature of their moral qualities, than on the power of their minds, and upon those very qualities which render them, in their appropriate sphere, the pride, the ornament, and the blessing of mankind . . . . [W]omen are distinguished for their passive fortitude . . . they have not that power of action, or active power of the mind which is essential to the practice of surgery. . . . [I]t is obvious that we cannot instruct women as we do men in the science of medicine.[37]

A third important factor is that physicians were in greater supply here than they were in Europe throughout most of the 19th century, and in greater supply in the years that were crucial to the midwife's survival—that is, until 1920—than they have been in recent years. This would have given economic competition more weight here than in Europe. The medical historian Shyrock notes:

> the ratio of doctors to population was far higher here than in Europe [in the period from 1860–1909], and competition was already keen when women began to enter it [medicine] . . . . [T]here was no little suspicion that doctors opposed women for fear of economic competition, the usual motive for suppressing a minority element. Occasionally this matter came out in the open, as when an article in the Boston Medical Journal noted as early as 1853 that competition was becoming serious: women were already cutting in on profits in obstetrical cases. In the same

city, Dr. Gregory, who suffered no inhibitions, accused the doctors of desiring a male monopoly for the market.[38]

In 1898, Henry Jacques Garrigues, an obstetrical surgeon to the Maternity Hospital in New York, wrote:

> Although an evil, midwives are, however, in most countries a necessity in view of the fact that physicians would be unable to find the time to do the work. This is not so here, where there is a superabundance of medical men.[39]

However, in 1921, when physicians were in much shorter supply, another physician describing a new program for inspecting and supervising midwives in Philadelphia made this interesting apology:

> It seems almost incredible that anyone, not a confirmed idealist, cognizant of conditions as they exist, will fail to realize without further argument that the most vital and compelling argument for considering the midwife a necessity today is that without her aid, it would be impossible to care for the cases of childbirth in any large city or other locality in which a large proportion of the population is composed of poor people. . . .
>
> We . . . resent the charge that we hold a brief for the midwife. On the contrary, we assert that she is an anachronism and a menace under the conditions at present operative, and that as soon as any practicable (emphasize practicable) scheme is evolved, that we will be the first to applaud her retirement to the oblivion which is her due.[40]

If indeed competition played a role in the midwife's disappearance, it would be logical to assume that powerful professional associations gave the physician an advantage in the eventual outcome. The role they played in blocking women from practicing medicine is described in Mary Roth Walsh's article. Just one example is added here:

> In 1859 the Medical Society of the County of Philadelphia passed "resolutions of excommunication" against every physician who should "teach in a medical school for women" and every one who should "consult with a woman physician or with a man teaching a woman medical student." In Massachusetts, after qualified women physicians were given state certificates to practice, the Massachusetts Medical Society forbade them membership, thus refusing to admit the legality of diplomas already sanctioned by the highest authority.[41]

The previously quoted Garrigues, a member of the prestigious New York Academy of Medicine, gives an example of the same sort of professional pressure being brought to bear on the activities of midwives. He states that in 1884 a bill was introduced into the state legislature to grant a charter to a certain college of midwives in the City:

> Dr. Albert Warden and I were sent to Albany as delegates to confer with persons of influence on the matter, and the bill was eventually killed.[42]

In 1898, he had himself offered a resolution at the Academy of Medicine, "To take immediate steps to pass a law confining the practice of midwifery to qualified medical practitioners. The resolution was passed January 1898."[43]

## Childbirth Care Today

Today, in the United States, there is renewed interest in training midwives, and graduate programs in midwifery are appearing in the more prestigious schools of nursing.[44] This trend coincides with a health care delivery crisis that is perceived and defined as a shortage of medical personnel,[45] and has given rise to the development of new programs to train physician's assistants, primary care nurses, and other paraprofessionals to lighten the physician's load. However, since the United States has more physicians per capita than many other developed countries with better health statistics, it seems likely that the "shortage" is due to distribution of physicians, an increased patient demand for services, as well as a continued reluctance of physicians to serve less affluent members of the population, and not to a lack of numbers.

Interestingly, women being trained to deliver babies are called *nurse*-midwives today, a clue to their status in the United States as paraprofessionals rather than independent practitioners. It appears likely that by and large the medical profession will continue to define and limit the scope of the nurse-midwife's field, and that she will perform under implicit if not explicit supervision of physicians.

Today, childbirth care is changing in several different directions. On the one hand, a slowly developing Birth Reform Movement, which has its roots in the work of Grantly Dick-Read over a generation ago, is bringing a safer, more personal, woman-and-infant oriented delivery to increasing numbers of women. On the other hand, concerned women and men, both lay and professional, have been working through such organizations as the International Childbirth Education Association, and the American Society for psychoprophylaxis in Obstetrics, to break the medical profession's definition of childbirth as pathological, and present it once again as a normal human event.[46] They have pioneered in educating women for unanesthetized (prepared) childbirth, and are bringing about some changes in the organization of hospital care.

However, these organizations essentially work within the system to reform it. They do not address the issue of present-day medical and male dominance of childbirth, which is perhaps more subtle than it was previously. Women are encouraged to participate in their own deliveries, but obstetricians still retain final authority to decide on intervention. And another male, the husband, is added, although it is rare indeed for a woman to be permitted another woman of her own choosing to be with her in labor. While it is important for a father to regain access to the birth of his own children and to be able to support the mother in labor, prepared childbirth methods frequently give him an authoritarian role he did not have before. In the Lamaze method, for example, he is taught to monitor the labor and instruct the woman in breathing. In the method taught by the American Academy of Husband-Coached Childbirth, he is given even more responsibility. Often he is cast in partnership with hospital personnel. In a popular film on Lamaze instruction, the father is shown conferring with the physician to decide whether to "permit" labor to continue or to use intervention to hasten it.

Leboyer's book *Birth Without Violence,* which has found wide appeal in its advocacy for gentler handling of babies after childbirth, provides another example of physician control. In describing his method for "non-violent" childbirth, the author ignores violence done to women by modern technological birthing. However, he describes in colorful language the imagined agony of the baby during labor. Not only the uterus but the woman herself is cast in the role of the one doing the violence—"the enemy" who "crushes and twists [the baby] in a refinement of cruelty"—a "monster" who "stands between the child and life."[47] After delivery, Leboyer disregards the mother and the mother-baby bond as he takes center stage, bathes the baby in warm water and caresses it with his gentle hands. It may be observed that his method seems "new" only in places where modern obstetrics has moved delivery into brightly lit, noisy, surgical operating rooms. In many other places (as in the past) babies are given warm baths, are gently handled by midwives and others, and are caressed by their mothers.

In discussing these attempts to reform hospital care, Anne Seiden has remarked that currently there is a "profound ambivalence about 'letting' the laboring woman control the situation, using companions of her choice including medical personnel as sources of support, consultation and technical expertise on an informed consent basis," and that there is "the potential indeed for a new and more sophisticated kind of infantilization of the pregnant or newly delivered woman."[48]

In the 1970s more radical movements have been seeking alternatives to the system, rather than reform. The Women's Health Movement is raising women's consciousness to the sexist orientation and poor quality of obstetric/gynecological care offered them. It is encouraging women to recognize the damage done them and their children by irresponsible hospital procedures; to learn about the functioning of their own bodies; and to assume control of their own health care, including childbirth.[49] In some areas, lay midwives and a few physicians who have rejected traditional hospital practices are assisting women in home deliveries.[50] However, these changes, as well as the more conservative reforms mentioned above, are affecting a minority of childbearing women, principally a self-selected population from the educated middle class.

Meanwhile, the majority of hospital deliveries by obstetrical specialists is moving toward *more* rather than less intervention. More sophisticated medication, anesthetics, and technology are used routinely, not only for the 5 per cent of deliveries that are complicated and in which intervention may save lives, but also for the 95 per cent in which it would be unnecessary, if sound prenatal and birthing care were available.[51] The number of deliveries by Caesarian section is increasing, running as high as 30 or 40 per cent in some areas. The risk to both mother and infant when intervention is used in otherwise normal deliveries has been well documented.[52] The higher rates of birth injuries, cerebral palsy, minimal brain damage, and infant death occurring in the United States than in countries where intervention is not the rule, leave little doubt that sophisticated technological birthing takes a grim toll. Unfortunately, this is still standard childbirth care for the great majority of American women, and certainly for the poor.

It remains to be seen what the Birth Reform Movement will mean for these women in the long run, and what part the "new" nurse-midwife may play in the story. Those who are working for change testify that it comes slowly and against great resistance from the medical profession.

## *Notes*

An earlier version of this essay has appeared in *Women and Health,* vol. 1, no. 6 (Nov./Dec./, 1976). Available from SUNY, College at Old Westbury, Westbury, L.I., NY 11568.

1. International Federation of Gynaecology and Obstetrics and the International Confederation of Midwives, *Maternity Care in the World: International Survey of Midwifery Practice and Training,* Report of a joint study group (Oxford: Pergamon, 1966), p. 13.
2. U.S. Department of Health, Education and Welfare, *Vital Statistics of the United States, Volume I—Natality* (Washington, D.C.: Gov. Printing Office, 1968, Table I-S, pp. 1–20.
3. International Federation of Gynaecology and Obstetrics, p. 3.
4. For a more detailed discussion *see* Mary Roth Walsh's essay in this collection, as well as her recent book, *Doctors Wanted: No Women Need Apply* (New Haven: Yale University Press, 1977).
5. European physicians, of course, had encountered the same problem establishing a claim superior to that of lay healers. But they could draw on the prestige of the Church, University and social class. For a discussion, *see* Barbara Ehrenreich and Deirdre English, *Witches, Midwives and Nurses: A History of Women Healers* (Old Westbury, NY: The Feminist Press, 1971).
6. Eliot Freidson, *Profession of Medicine* (New York: Dodd, Mead, 1970), p. 17.
7. Niles Newton and Margaret Mead, "Cultural Patterns of Perinatal Behavior," *Childbearing: Its Social and Psychological Aspects,* Stephen A. Richardson and Alan Guttmacher, eds. (Baltimore: Williams and Wilkins Co., 1967), pp. 193–194.
8. *See Education for Nurse Midwifery* (New York: The Maternity Center Association, 1967).
9. For a discussion of the physician's role *see* Talcott Parsons, *The Social System* (New York: Free Press, 1951), pp. 428–97.
10. U.S. Department of Health, Education and Welfare, *Vital Statistics,* pp. 1–20. Moving childbirth to the hospital has *not* meant safer deliveries for women and babies. By comparison, in the Netherlands in the late 1960s, over half of all births occurred at home with the assistance of a midwife and a maternity aide [Doris Haire, *The Cultural Warping of Childbirth* (Milwaukee: The International Childbirth Education Association, 1972), p. 15.] Yet in that country, the infant mortality rate for 1970 was 12.7 compared to 19.8 in the United States, which had higher infant death rates than fourteen other developed countries (United Nations Statistical Office).
11. William R. Rosengren and Spencer De Vault, "The Sociology of Time and Space in an Obstetrical Hospital," *The Hospital in Modern*

*Society,* Eliot Freidson, ed. (New York: Free Press, 1963), pp. 266–92.

12. Doris Haire, p. 12. For a recent comparative study showing lower rates of complications in home births compared to obstetrical hospital managed deliveries *see* Lewis E. Mehl, M.D., "Options in Maternity Care," *Women and Health,* September/October 1977, pp. 29–43. For a discussion of increased pathology in childbirth as a consequence of technological intervention in obstetrical management, *see* Frederick M. Ettner, M.D., "Hospital Technology Breeds Pathology," pp. 17–23, in the same issue of *Women and Health.*

13. Eliot Freidson, *Professional Dominance* (New York: Atherton, 1970), p. 277.

14. Episiotomy is the surgical cutting and suturing of the birth canal, ostensibly to prevent its "tearing" during delivery. For a discussion of the widespread prevalence of this and other surgical procedures in United States obstetrics, see Rosengren and De Vault cited above. For a discussion of the greater risk to mother and baby of surgical intervention in normal childbirth, *see* Haire; Suzanne Arms, *Immaculate Deception* (New York: Houghton Mifflin, 1975); and Nancy Stoller Shaw, *Forced Labor: Maternity Care in the United States* (New York: Pergamon Press, 1974).

15. Newton and Mead, *Childbearing,* pp. 171–72.

16. A. M. Carr-Saunders and P. A. Wilson, *The Professions* (Oxford: Oxford University Press, 1933), pp. 121–25.

17. Ibid. *See also* Ehrenreich and English's discussion of the Church's persecution of women healers, and its protection of physicians.

18. Herbert K. Thoms, *Chapters in American Obstetrics* (Springfield, Ill.: Charles C. Thomas Press), reprinted from *Yale Journal of Biology and Medicine* (1933), pp. 11–13.

19. Elizabeth Nihill, Professed Midwife, *A Treatise on the Art of Midwifery* (London: A. Morley, 1760), pp. 5–6.

20. Sarah Stone, *A Complete Practice of Midwifery* (London: T. Cooper, 1737), p. 16. *See also* the discussion of men-midwives and the barber surgeon guilds in Carr-Saunders and Wilson, p. 122.

21. Thoms, *Chapters,* p. 13.

22. *Ibid.,* p. 13.

23. *See* for example Nihill, *Treatise,* p. 9, and Stone, *Complete Practice,* p. 12.

24. For example, a London physician named John Stevens published a pamphlet in 1830 with the title *An Important Address to Wives and Mothers on the Dangers and Immorality of Man-Midwifery.* In it, he says: "We see in the works on midwifery, especially those which profess to trace the art to its origin, that surprise is expressed how *few* accidents would happen in former times when none but females assisted during parturition. The City of London Hospital

affords a most powerful proof of what female intelligence can perform" (pp. 29–30).

25. Thoms, *Chapters*, p 21. *See also* Samuel Bard, M.D., *A Compendium on the Theory and Practice of Midwifery* (1807). Dr. Bard was the president of the College of Physicians and Surgeons in the University of the State of New York, and his text, which went into several editions, was the first text on obstetrics published here.

26. For example *see* William Felix Mengert, M.D., "The Origin of the Male Midwife," *The Annals of Medical History* 4, 5 (1932): pp. 453–65. This article is a reprint of a work originally published in mid-19th century.

27. George Gregory. Published in New York, 1853.

28. John Stevens, M.D. Published in London, 1830.

29. Pamphlet, 1851.

30. Freidson, *Dominance*, p. 21.

31. George Rosen, "The Hospital: Historical Sociology of a Community Institution," *The Hospital in Modern Society*, Eliot Friedson, ed. (New York: Free Press, 1963), p. 25.

32. United States Bureau of the Census, *Statistical Abstract of the United States* (Washington DC, 1970), p. 45. *See also* Carolyn Conant van Blarcomb, "Midwives in America," *American Journal of Public Health* 8, 4 (1914): 197–207.

33. *Ibid.,* p. 198.

34. For a discussion of the issues, and a report of Parliamentary hearings leading to the passage of the act, *see* British Parliamentary Papers, *Report of the Select Committee on Midwives Registration Together with the Proceedings of the Committee,* 17 June 1892. Ordered by the House of Commons General Health Session 1890–1894.

35. Cynthia Epstein, *Woman's Place* (Berkeley: University of California Press, 1971), p. 41.

36. *See* Mary Roth Walsh's paper in this collection, as well as her book, *Doctors Wanted: No Women Need Apply.*

37. A Physician, *Remarks on the Employment of Females as Practitioners in Midwifery* (Boston, 1820).

38. Richard Harrison Shyrock, *Medicine in America, Historical Essays* (Baltimore: Johns Hopkins Press, 1966), p. 187.

39. Henry Jacques Garrigues, "Midwives," *Medical News* (Philadelphia), 19 February 1898, p. 98.

40. A. B. Nicholson, M.D., "The Midwife, an Anachronism of the Twentieth Century," Lecture delivered in Philadelphia, 1921, pp. 10–11 (Reprint in New York Academy of Medicine Library).

41. Anna Garlin Spencer, "Woman's Share in Social Culture," *Feminism: The Essential Historical Writings,* Miriam Schneir, ed. (New York: Vintage, 1972), pp. 269–85. Quote from p. 282.

42. Garrigues, *Midwives*, p. 11.

43. *Ibid.,* p. 12.
44. John Kosa, "Women and Medicine in a Changing World," *The Professional Woman,* Athena Theodore, ed. (Cambridge: Schenkman, 1971), pp. 709–20. *See also, Education for Nurse Midwifery.*
45. Patricia L. Kendall and George G. Reader, "Contributions of Sociology to Medicine" *Handbook of Medical Sociology,* 2nd ed., Howard E. Freeman, Sol Levine, and Leo G. Reeder, eds. (Englewood Cliffs: Prentice-Hall, 1972), p. 9.
46. Anne M. Seiden, "The Birth Reform Movement: Strengths and Some Limitations" (Paper delivered at the Third Annual Conference on Psychosomatic Obstetrics and Gynecology, held at Temple University, Philadelphia, Pa., February 2, 1975).
47. Frederick Leboyer, *Birth Without Violence* (New York: Knopf, 1975).
48. Anne M. Seiden, *Birth Reform,* p. 5.
49. *Proceedings of the First Childbirth Conference* (Stamford, Ct.: New Moon Communications, 1973).
50. Raven Lang, ed., *The Birth Book* (Ben Lamond, CA: Genesis Press, 1972); and Suzanne Arms, *Immaculate Deception.*
51. Anne M. Seiden, "The Sense of Mastery in the Childbirth Experience," *Primary Care* 3, 4 (1976), 717–25.
52. Doris Haire, *Cultural Warping,* p. 12.

[Datha Clapper Brack *is Assistant Professor of Sociology at Bergen Community College, Bergen, New Jersey, where she teaches courses on Sociology of Sex Roles and on Society and Women's Health. She received her Bachelor and Master degrees in Sociology from Columbia University, and is currently completing her dissertation in Medical Sociology at C.U.N.Y. While specializing in women's health issues, she is presently conducting research on social forces influencing women to choose breastfeeding and to breastfeed successfully. Some of her preliminary research on this topic has been published in* Nursing Outlook. *She is a member of Bergen Community College's Affirmative Action Committee, and the national and local chapters of Sociologists for Women in Society. She writes: "Bearing and rearing six children in suburbia probably did more to raise my feminist consciousness and push me to study women's health issues than any other single experience."*]

Figure XIII. Boston University Medical Center, class of 1886. (Boston University School of Medicine; collection of Mary R. Walsh)

# Mary Roth Walsh

## The Quirls of a Woman's Brain

That there was a substantial number of women physicians in the nineteenth century comes as a surprise to many. For example, Dr. Charles Phelps recently wrote in the *Journal of Medical Education* that the increased number of women medical students "represents a rather remarkable change considering the fact that there were no women in medicine only a hundred years ago." In reality, the late nineteenth century witnessed a sharp upsurge in the numbers of women physicians. By 1890, only forty-three years after the admission of the first woman, Elizabeth Blackwell, to medical school, there were 4,557 women doctors in the United States. In the same year, 18 percent of Boston's physicians were women, a remarkable number when one considers that in 1976, women accounted for only 11.7 percent of Boston's medical population.[1]

Women's success in gaining access to medical schools in the late nineteenth century seemed to guarantee that the number of female physicians would continue to mount. In 1893, women made up 10 percent or more of the student enrollment at eighteen regular medical schools across the nation. In the same year, women accounted for 19 percent of the University of Michigan Medical School and 31 percent of Kansas Medical College. These figures, of course, must be balanced against cities like New York, Philadelphia, and Chicago, where all of the regular medical schools were still sex segregated, clustering the women students in women's medical colleges. In other cities, women found themselves welcome only at the irregular medical colleges, particularly the homeopathic schools. (See Table I for the names of some of the long established regular medical schools that admitted 10 percent or more women in 1893–94.)

Moreover, the evidence of women doctors' success in the nineteenth century is not limited to their numbers. Articles in the press measured their success with traditional financial yardsticks as well. *The Boston Daily Advertiser* claimed that scores of women doctors, who counted among their patients the city's "most cultivated, influential, and high

*Table 1*

*Regular Medical Schools with 10 percent enrollment of female students in 1893–94*

| College and Location | 1893–1894 Total Enroll. | % Women |
|---|---|---|
| 1. Univ. Southern Calif. | 39 | 15.38 |
| 2. Cooper Med. College (Calif.) | 228 | 12.28 |
| 3. Univ. of Calif. | 109 | 11.01 |
| 4. Med. Dept., Univ. Col. | 42 | 19.05 |
| 5. Denver Medical Coll. (Col.) | 36 | 25.00 |
| 6. Gross Medical Coll. (Col.) | 72 | 35.85 |
| 7. National University (D.C.) | 88 | 12.50 |
| 8. Council Bluffs Med. Coll. (Iowa) | 12 | 25.00 |
| 9. Kansas Medical College | 44 | 31.11 |
| 10. Coll. Phy. and Surgeons (Mass.) | 135 | 21.48 |
| 11. Tufts University (Mass.) | 80 | 28.75 |
| 12. Johns Hopkins Univ. (Maryland) | 83 | 15.66 |
| 13. Univ. of Michigan | 375 | 18.93 |
| 14. University of Buffalo | 186 | 11.83 |
| 15. Syracuse Univ. (New York) | 49 | 14.29 |
| 16. National Normal Univ. (Ohio) | 26 | 13.33 |
| 17. Toledo Medical College (Ohio) | 38 | 10.53 |
| 18. Univ. of Oregon | 29 | 17.24 |

These figures are drawn from figures given in The U.S. Commissioners of Education Report.

born women," had incomes of five figures. Another article in the *New York Herald Tribune* singled out the success of Boston women physicians and called attention to the "surprising number of Back Bay offices luxuriant in appointment of tasteful furniture, paintings, and bric-a-brac belonging to women who add M.D. to their names." The *Boston Post* cited several Boston University Medical School graduates "whose practice is lucrative" and whose "professional services are in demand in some of the best families of the city."[2]

Novelists of the day quickly took note of this new phenomenon and women physicians made their appearance in William Dean Howells' *Dr. Breen's Practice* (1881), Sarah Orne Jewett's *A Country Doctor* (1884), Elizabeth Stuart Phelps' *Dr. Zay* (1886), Henry James's *The Bostonians* (1886), and Annie Nathan Meyer's *Helen Brent, M.D.* (1891).[3]

A great deal of the success of nineteenth-century women physicians was due to the support of the feminist movement.[4] Women raised funds to help other women gain a medical education. They also helped finance hospitals such as the New York Infirmary and the New England Hospital for Women and Children, institutions with all-female staffs run by women for women.

These medical bastions not only symbolized female solidarity, they became centers where women physicians could expand their medical base. As we shall see shortly, those who wished to keep women out of medicine often based their arguments on the physical inferiority of women. Female physicians in women's hospitals were able to counter these attacks with research that demonstrated their scientific as well as human capacity for practicing medicine.

Another major source of encouragement was the feminist press. Lucy Stone's *Woman's Journal* regularly attacked the barriers to female careers in medicine and urged its readers to patronize female physicians. A dramatic illustration of the commitment of the feminist movement to the professional advancement of women was its push to open the elite medical schools to women. Although more and more medical colleges began to admit women in the last quarter of the nineteenth century, a

*Figure XIV.   Dr. Anna Howard Shaw and other suffragists.*
*(Collection of Mary R. Walsh)*

number of women believed that matriculation at a prestigious school
would insure their success. Their strategy was simple: to find an elite
medical school in need of funds and offer a gift of money in return for
the beneficiary's promise to admit women. In 1873, Dr. Mary Putnam
Jacobi, one of America's leading physicians, commented on her own
purchasing of privileges while a student at the École de Medicine in
Paris: "It is astonishing how many invincible objections on the score
of feasibility, modesty, propriety, and prejudice will melt away before
the charmed touch of a few thousand dollars."[5]

But the women soon discovered that it would take more than a few
thousand dollars to buy their way into a major medical school. In 1878,
Harvard Medical School rejected feminist Marion Hovey's offer of
$10,000, if the school would agree to educate women on equal terms with
men. Three years later, a group of women physicians in Boston raised
$50,000 in pledges from women in several cities. At first, Harvard
accepted the offer, but quickly reversed itself when an enraged medical
school faculty threatened to resign if women were admitted as students.[6]

Undaunted by the collapse of the Harvard venture, women shifted their
campaign to an even more attractive possibility, Johns Hopkins Medical
School. The financially beleaguered Baltimore school had long been
in the planning stage, but it had been unable to open because of lack of
funds. Convinced that Johns Hopkins offered the best hope, a number
of women collected $500,000 by 1892. They demanded that Johns
Hopkins admit women on the same terms as men, *and* require an A.B.
degree for admission, a prerequisite that no other medical school in the
country imposed. The administration had no choice but to accept
the women's terms and, in 1893, three out of the twenty-one entering
students in the first class were women.

The victory over Johns Hopkins appeared to signal the end of
women's second-class status in medicine. The *Nation* echoed the hopes
of many when it enthusiastically proclaimed that John Hopkins could be
the turning point for women physicians. *Century* magazine marked the
decision by publishing a series of letters of approbation from a diverse
group of public figures that included M. Carey Thomas, the president
of Bryn Mawr College, Dr. Mary Putnam Jacobi, Josephine Lowell,
and Cardinal Gibbons. What is remarkable, the *Nation* noted, is that not
a single one of these writers had "a word to say on the question of
whether the quirls of a woman's brain have any peculiarities which
necessarily unfit her from profiting from the most advanced medical
instruction."[7]

The "quirls of a woman's brain" had been an issue, however, for the
last half of the nineteenth century. Moreover, the *Nation* was naive in

assuming that the Johns Hopkins capitulation marked the final surrender of those who saw woman's biology as an insurmountable barrier to professional achievement.

The efforts of women to secure equal opportunity in medicine had, of course, encountered determined opposition from the male medical establishment. The arguments of the more articulate medical men formed not only a major obstacle to women's advancement in medicine, but also furnished the nineteenth-century antifeminists with a rich source of ammunition. And while historians have recently begun to examine the scientific and medical rhetoric about appropriate sexual spheres, the highly charged situations that spawned the rhetoric have been ignored.

During the last half of the nineteenth century, the age-old question of woman's place became a central issue. Women were knocking on doors marked "men only" and demanding to be admitted. Medical men found themselves confronted with some of the same problems that their friends in other fields faced, but felt themselves particularly vulnerable to a female onslaught. In colonial times, women had dominated midwifery,[8] and if women were to enter any profession, their special "talent" for nurturing seemed to dictate a career in medicine.

Unfortunately, American women began to press their efforts to be admitted to medical schools at a time when physicians were concerned about the depressed state of the profession, a condition that many believed stemmed from an oversupply of doctors. Moreover, women made up a large portion of many a medical man's practice; what would happen if women heeded the feminist press's call to support their sisters who were struggling to establish themselves? One solution was to prove that a woman's nature, far from being an asset in a medical career, was an insurmountable liability. Nowhere in the profession was there a greater urgency to promote this idea than among those men who specialized in gynecology and obstetrics, the areas where women physicians posed the greatest threat.

What resulted was the most extensive emotional debate on women's biology in American history. In 1866, the *Boston Medical and Surgical Journal* declared that the issue of women physicians was creating chaos in the profession and now was the time to bring the "unsettled question" into the open. As if waiting in the hospital wings, Dr. Horatio Storer quickly responded to the challenge.

Storer was a complex figure in Boston medicine. The son of David Humphreys Storer, Head of the Obstetrical Department at Harvard, Horatio had pioneered in separating the study of gynecology from obstetrics, and was one of the founders and, later, president of the

*Figure XV.   Dr. Horatio Robinson Storer*
*(Countway Library of Harvard Medical School; collection of*
*Mary R. Walsh)*

Gynecological Society of Boston. In 1863, he had joined the New
England Hospital for Women and Children as head surgeon, the only
male to serve on the hospital staff throughout the nineteenth century.

Storer's response to the *Boston Medical & Surgical Journal* came a
few months after the challenge was printed, in his letter of resignation
from the staff of the Women's Hospital. The letter, which focused on
women's unsuitability for a career in medicine, was all that the medical
journal could have hoped for.[9] And what better evidence could there be?
Here was a man who had observed female physicians, not from afar,
but alongside them in the wards and operating rooms of their own
hospital. Furthermore, he had been assisted for two years in his private

practice in Boston by Dr. Anita Tyng, whom he described as "one of the very best woman physicians . . . as I suppose there is at present in the country . . . [whose] natural tastes and inclinations . . . fit her, more than I should have supposed any woman could have become fitted, for the anxiety, the nervous strain, and shocks of the practice of surgery." Nevertheless, she was dismissed by Storer as an "exception," for women were "naturally" lacking the courage and the daring to pursue the dangerous and difficult decisions involved in gynecological surgery.

But what else could one expect? How could women act freely and confidently when they were the captives of their own biology? Here, Storer turned to the subject to which every medical opponent of women was irresistibly drawn in the nineteenth century—the female reproductive system. Storer asked rhetorically, who could trust the great questions of life and death to women whose equilibrium varied from "month to month and week to week . . . up and down?" It was not to women as physicians, *per se,* that Storer objected, for he claimed they made "most agreeable and charming attendants." What Storer found objectionable was "their often infirmity during which neither life nor limb submitted to them would be as safe as at other times." It was clear from Storer's description of menstruation as "periodical infirmity . . . mental influences . . . temporary insanity," that he believed women to be monthly cripples, certainly more in need of medical aid than able to furnish it. But if women remained in their proper sphere, all would be right with the world. Storer's views of the proper sphere for that crippled class of individuals coincided with those of another widely quoted physician of the era, Dr. Charles Meigs, who wrote in his textbook on obstetrics that woman "has a head almost too small for intellect but just big enough for love."[10]

Storer's opinion was very persuasive to those who wished to be persuaded. As a leading physician, a man of obvious good will who had worked in a women's hospital and so, in a sense, had risked his reputation to conduct a scientific experiment in the opposition's laboratory, Storer's pronouncement that the experiment had proven to be a failure was not taken lightly. Yet the question must be asked, were the experiment and Storer's findings as objective as Storer intimated?

For Storer, 1866 had certainly been a difficult year. His disagreement with senior members on the Harvard Medical Faculty had led to his dismissal in the spring of that year from the position he had held as assistant in obstetrics, and had ended permanently his connection with the Harvard Medical School. And tucked away in his letter of resignation to the New England Hospital was a brief reference to his immediate reason for leaving: his objection to a new requirement that surgeons must consult with their colleagues before performing high risk surgery. On 13 August 1866, the Board of Directors of the New England Hospital had joined forces to prevent further danger to women patients and put forth the following declaration:

> Whereas: The Confidence of the Public in the Management of the Hospital rests not only on the character of the Medical attendants, having its immediate charge, but also on the high reputation of the consulting physicians and surgeons, and, Whereas: We cannot allow them to be responsible for cases over which they have no control—
> Resolved: That in all unusual or difficult cases in medicine, or where a capital operation in surgery is proposed, the attending and Resident Physicians and Surgeons shall hold mutual consultations, and if any one of them shall doubt as to the propriety of the proposed treatment or operation one or more of the consulting physicians or surgeons shall be invited to examine and decide upon the case.[11]

The Board's action was in response to the fact that all three patient deaths during the previous year had occurred in the surgical wards of the hospital, after what Dr. Lucy Sewall described as "hazardous operations." As Dr. Mary Putnam Jacobi, who had been an intern at the hospital while Storer was on the staff, later recalled, the results of Storer's operations often failed to match the boldness of his plans. And in 1866, the hospital's action came as part of a double blow, following Storer's recent dismissal from Harvard. In addition, the requirement that he clear his operations with female physicians put Storer in what he regarded as a subordinate position. Of even greater significance was the fact that these restrictions blocked what had been the hospital's major attraction for Storer—a free hand to develop his skills as a gynecological surgeon.[12]

In 1863, when Storer had first joined the hospital, none of the other hospitals in Boston allowed abdominal gynecological surgery. In fact, the entire field of gynecology was treated with great suspicion by the conservative Boston medical establishment. Storer had even been warned that the profession in New England would never tolerate in its ranks an "avowed gynecologist." Storer's son, Malcolm, who also became a physician, later attributed the prejudice regarding gynecology to the low status associated with the treatment of women's diseases: "In the ears of conservative men, the very name of diseases of women savored strongly of quackery; and it was the honest belief of many a doctor of the old school that the preservation of a man's personal morality was highly dubious if he was constantly engaged in treating the female genitals . . ."[13]

Moreover, the few experiments in gynecological surgery in Boston had all proven to be failures. Six women had been operated on for ovarian tumors between 1830 and 1858 at Massachusetts General Hospital; but, after all six women died, gynecological surgery was not permitted inside the hospital until asepsis (modern sterilization procedure) was fully established in the 1880s. The other major Boston hospital, Boston City, had similar strictures against such operations, a fact not surprising since the medical leadership in Boston was a closely knit group that oversaw all hospital affairs. That the women's hospital in Boston permitted surgery out of concern for women patients who might be saved by an operation is testimony to the women doctor's confidence in their own procedures. Other evidence indicates that many leading male physicians in the city shared their faith in the hospital's procedures.[14]

Although Storer was later able to continue his work in the newly opened, but less prestigious, operating rooms of the Carney and St. Elizabeth's hospitals, the increasingly conservative medical atmosphere in Boston limited the extent of his experimentation and publications. Nevertheless, Storer's later operations continued to exhibit much of the "boldness" to which the New England Hospital had objected. Male surgeons have historically taken a cavalier view of operations on the female reproductive system, in sharp contrast to their reluctance to experiment on the male organs. Thus, a later physician turned historian could describe a three-hour gynecological operation by Storer in 1868 as "the greatest feat" in his career. Similarly, Storer's assistant in the operation depicted it as "the most heroic of the bold precedures as yet resorted to." Both observers, however, glossed over the fact that the woman died![15]

What made Storer's lost or limited opportunities especially frustrating was the knowledge that free-wheeling gynecological surgery was being performed elsewhere in the nation. During the late nineteenth century, Storer was easily outdistanced by his daring medical rival in New York, Dr. J. Marion Sims, who performed his operations in crowded amphitheatres. In one marathon display of his surgical virtuosity, Sims performed a series of varied operations for four successive days, capped off by an entertainment dinner for the large audience of distinguished American doctors.

Furthermore, whereas Storer had been thwarted by the New England Hospital women, Sims had vanquished the wealthy society women who had founded the New York Hospital for Women in 1856. Sims had been appointed head surgeon with the expectation that he would engage a woman as his assistant, hopefully Dr. Emily Blackwell, who had just returned from study abroad and was eminently qualified. Sims at first resisted, but when the women insisted, he derisively appointed a female acquaintance who had been serving at the hospital as matron and general superintendent and who had no medical training whatsoever. Six months later, the board of lady managers of the hospital backed down and a man was appointed to assist Sims in his surgery, the selection being based not on medical qualifications, but on the fact that the man had married a young Southern friend of Sims' acquaintance.[16]

Sims had apparently recognized that female physicians might serve as a check on his aspirations. Unhampered by the type of resistance Storer had encountered at the New England Hospital, Sims performed an incredible variety of gynecological operations in the following years. Having previously performed thirty experimental operations on a slave woman named Anarchia, he now performed a similar number on an Irish woman, Mary Smith, in New York. His slashing scalpel dazzled the medical world, and earned him the reputation of "one of the immortals" in gynecological surgery. Among the medical students at Harvard, he was recognized as possessing "divinity."[17]

Physicians such as Storer and Sims, who were engaged in gynecology and obstetrics, were especially apprehensive about women in medicine. Because of the sexual anxiety of Victorian culture, the increasing numbers of women physicians threatened a revival of the old charges of immorality and insensitivity laid against men when male midwives first began to practice. In the seventeenth and eighteenth centuries, the "man-midwives" had to struggle to gain acceptance from their female patients and were much criticized by woman-midwives and looked down upon by doctors. That this criticism spilled over to the field of gynecology in the nineteenth century is demonstrated by the fact that at the first meeting of the Gynecological Society of Boston, founded by Storer in 1869, a resolution was passed confirming that the male physician who treated female patients exclusively was not affected by lust and sensuality. The Society declared "all impurity of thought and even the mental appreciation of a difference in sex is lost by the physician."[18]

Storer's career demonstrates that a good deal of the pseudo-scientific opposition to women in medicine stemmed from what can hardly be described as dispassionate sources. Far from the picture that they sought to project of men passing judgments from a vantage point above the battle, the view of male physicians was often shaped by their own special conflicts with women in medicine. We do know that when Horatio Storer was finally forced to withdraw from active medical practice in

1872, because of a near fatal infection from an accident in one of his hazardous operations, the opponents of women in medicine lost a vigorous spokesman.

It is testimony to the growing intensity of the struggle that his place was taken by an even more formidable figure, one who became a national spokesman for the anti-feminists in America, Dr. Edward Clarke. Clarke had succeeded the distinguished Dr. Jacob Bigelow in 1855 as professor of *materia medica* at the Harvard Medical School, but resigned from the Harvard faculty in 1872 and became a member of Harvard's Board of Overseers. Although Clarke was in no way a colleague of Storer's— he had, in fact, been one of those on the faculty responsible for the gynecologist's dismissal from Harvard in 1866—the two men shared similar views on the need for a masculine and "scientific" examination of the woman question.

Like Storer, Clarke had been a liberal for several years on the question of women's advancement. Early on, in 1869, Clarke had felt obligated to respond to the growing animosity shown by the male medical students in Philadelphia who had driven the women students out of the classroom with a shower of tobacco quids and tinfoil.[19] In an article published in the *Boston Medical and Surgical Journal,* he described the Philadelphia incident as "unfortunate" because of both the "unenviable notoriety" brought upon the young men involved and the fact that "nothing advances any cause so much as the martyrdom or persecution of its disciples. In this way the Philadelphia medical class have given an unexpected impetus to the cause they opposed." And indeed, this is exactly what happened. That very same year, almost directly linked to the incident, the Philadelphia County Medical Society proposed a resolution for the admission of women. Clarke apparently hoped to ward off the brash activities of immature young medical students, but also any premature action by his medical brethren. He hoped that the publication of his remarks, originally presented to the graduating class the previous spring at Harvard, would contribute to a discussion of the subject of women physicians "till a satisfactory solution is reached."[20] There is little doubt that Clarke hoped that without any catalytic incidents such as had occurred in Philadelphia, the discussion would proceed at a leisurely pace.

In his essay he reminded his audience, who probably needed no reminding, that women were knocking hard at the physician's door. Pointing out that whatever woman can do, she has a right to do and eventually *will* do, Clarke stated that *a priori,* she had "the same right to every function and opportunity which our planet offers, that man has." Neither did Clarke believe that there was anything in medicine that was improper for women to study—"for science . . . may ennoble, it can never degrade man, woman, or angel." But, he warned, the real question was not one of right, but of capability. If woman were capable, no law, argument, or ridicule would prevent her success. Therefore, Clarke urged, neither the medical profession nor the community should stand in her way: "Let the experiment . . . be fairly made . . . [and] in 50 years we shall get the answer." Clarke volunteered to predict that women could master the science of medicine.[21]

The address was certainly a welcome contrast to Storer's repeated declaration that the experiment had in fact already been tried and proven a failure. There were, however, a few items in Clarke's talk that should have set off warning signals in the feminist camp. Although he had stated that women could master medicine, he did not define what he meant by this. Nor did he explain what he meant when he declared that, with the exception of a few areas, he felt women would not become successful medical practitioners. But what difference did such comments make? All the women physicians wanted was the fair test Clarke had sanctioned. They were convinced that they would pass with flying colors.

More menacing was Clarke's insistence that the test should take place in separate classrooms. Although Clarke felt that there was nothing improper in the study of medicine by women, he seemed to believe that something mysterious and dangerous would materialize when the pursuit took place in a coeducational context. After all, a bath was a necessary and even purifying process for all, claimed Clarke, but he warned that it did not follow that the two sexes needed to bathe "at the same time and in the same tub." No narrow-minded reactionary, Clark welcomed the suffrage for women but "God forbid that I should ever see men and women aiding each other to display with the scalpel the secrets of the reproductive system; . . . or charmingly discuss together the labyrinthine ways of syphilis."[22]

But the most ominous note was struck by Clarke when he outlined the types of questions that should be included in the test of woman's fitness to be a physician. The heart of the matter was the issue of whether or not woman's nature would enable her to advance in medicine. He claimed the answer in the final analysis would be found in woman's physiology—"the facts which physicians can best supply." Nevertheless, in spite of its promises of trouble, Clarke's address did not touch off any public criticism from the women, who apparently chose to concentrate on the positive notes in his talk.

Women, quite clearly, were both eager and able to participate in Dr. Clarke's 50-year experiment. In fact, they were too eager as it turned out

—at least, as far as Clarke himself was concerned. It is difficult to explain his sudden reversal and the disappearance of his liberality with regard to the advancement of women. It may have been due to the fact that women were now pressing Harvard itself to open its doors. And Clarke, one assumes, was elected to the Board of Overseers to guard Harvard's male sanctity. It was one thing to talk nobly of a half-century's test of time; it was something quite different to look out and see contestants preparing to storm the gates of one's own home. Clarke, who had received his M.D. at the University of Pennsylvania, loved Harvard with the zeal of a convert.

He spoke to all these issues—the immediacy of the women's petitions to enter Harvard, the proper role of women, and the sacredness of an all-male educational environment—at the New England Women's Club of Boston in December, 1872. Although his own medical specialty was otology, the study of the ear, he picked up the challenge he had thrown out in 1869 and proceeded to discuss the relationship between women's education and their physiology. Not surprisingly, he found women limited by their biology. Clarke was surprised by the furor that erupted after his talk in Boston and decided to review his statements carefully with an eye to publishing them in a more comprehensive form.

The result was *Sex in Education; or, A Fair Chance for the Girls,* published in the fall of 1873. Perhaps no other single book on the limitations of the female system evoked such a wave of controversy. Within thirteen years, Clarke's book went through seventeen editions. As far away as Ann Arbor, Michigan, it was reported that everyone was reading Dr. Clarke's book. A local bookseller there claimed sales of two hundred copies in a single day, chortling, "the book bids fair to nip coeducation in the bud." But neither the number of printings nor their geographical distribution reflects the full impact of the book. Years later, M. Carey Thomas, the first President of Bryn Mawr College, recalled that "we did not know when we began whether women's health could stand the strain of education. We were haunted in those days, by the clanging chains of that gloomy little specter, Dr. Edward H. Clarke's *Sex in Education.*"[23]

Whereas Clarke in his 1869 address at Harvard had originally opposed coeducation only in the sensitive area of medical instruction, in his 1872 speech, he reworked and expanded his thesis to encompass all post-puberty education. The most dangerous threat, Clarke believed, came from the mistake of educating females as if they were males. Clarke argued that unlike the male whose development into manhood he viewed as a continuous growth process, the female at puberty experienced a sudden and unique spurt during the development of her reproductive system. If this did not occur at puberty, or if some outside force interfered, this "delicate and extensive mechanism within the organism,— a house within a house, an engine within an engine" would fail to develop. Since he believed the uterus was connected to the central nervous system, energy expended in one area (i.e. the growth of one's mind) was necessarily removed from another (i.e. the growth of one's uterus).

In *Sex in Education,* Clarke cited the dramatic case of Miss D_____ who entered Vassar College at the age of 14, a normal and healthy girl. Within a year, menstruation began and Miss D_____ continued to follow what Clarke described as the normal regimen of a male student. The results were predictable. Fainting was followed by painful menstruation, but she persisted in her studies until at last she graduated with "fair honors and a poor physique." In the following year, the young woman was "tortured" for two or three days out of every month and left weak and miserable for several more days. Then, the flow stopped altogether and Miss D_____ became pale, nervous, hysterical and complained of constant headaches. On examining the girl, Clarke found evidence of arrested development of the reproductive system. Confirmatory proof was found in his examination of her breasts, "where the milliner had supplied the organs Nature should have grown."[24]

In the pages that followed, Clarke went on to describe six similar cases in which the women all experienced "those grievous maladies which torture a woman's earthly existence: leucorrhoea, amenorrhea, dysmenorrhoea, chronic and acute ovaritis, prolapsus uteri, hysteria, neuralgia, and the like." And, if this were not enough, Clarke painted the end results of female education: "monstrous brains and puny bodies; abnormally active cerebration and abnormally weak digestion; flowing thought and constipated bowels. . . ."[25] The wonder is not that Dr. Clarke's book loomed so large in the thinking of women like M. Carey Thomas, but that they dared to entertain any thought of education at all.

The implications Clarke drew went far beyond the field of education; they extended into the sphere of population problems. The increasing number of educated women would mean that within 50 years "the wives who are to be mothers in our republic must be drawn from trans-Atlantic homes."[26] Clarke's study dovetailed neatly with the growing concern over the shrinking size of the American family—especially among the genteel classes. As early as 1850, the Massachusetts Census returns indicated that the foreign-born had a considerably higher birth rate than most white native Americans, a situation that would later give rise to fears of race suicide.[27]

Once again, women found themselves blocked by someone who had originally appeared to be a friend. Five years after urging a fair test of women's capabilities, Clarke declared his dread of seeing "the costly experiment" tried. His only solution was to provide women with *"a special and appropriate education, that shall produce a just and harmonious development of every part."*[28] Clarke was vague about the particulars of this special education, except to recommend that girls spend only two-thirds the time boys spent on their studies, and that the girls be given time off during their menstrual periods.

The seriousness with which Clarke's opponents treated his work can be measured by the extent of their response, which included at least four books published in 1874 alone: *Sex and Education; No Sex in Education; Woman's Education and Woman's Health;* and *The Education of American Girls.* Each critic recognized the dangers if Clarke's book remained unchallenged. As Mrs. E. B. Duffey, editor of *No Sex in Education,* noted, Clarke's covert plan "has been a crafty one and his line of attack masterly. He knows if he succeeds . . . and convinces the world that woman is a 'sexual' creature alone, subject to and ruled by 'periodic tides,' the battle is won for those who oppose the advancement of women."[29]

The opposition hammered away at each of Clarke's points. His critics agreed that his study would fail any scientific test. As Thomas Wentworth Higginson pointed out, to take seven cases out of a physician's notebook, assuring the readers that there were a good many more, was simply not good enough. Furthermore, Clarke neglected to present seven "representative" males for comparison, but simply assumed that boys could withstand any educational pressures to which they were exposed. Similarly, the resident physician at Vassar questioned Clarke's integrity in selecting the seven women, noting that the case of Miss D_____ was not even possible since no girl as young as 14 had ever enrolled at the college. She claimed that an error of such proportion could not help but shake one's confidence in the book's other cases, and, indeed, in its very thesis.[30]

Critics generally agreed that whatever difficulties were experienced by female college students were environmentally induced. Julia Ward Howe asserted that, if anything, a woman's education should be more like a man's, in that she should be given equal amounts of exercise and fresh air rather than be confined to the home after school to perform her domestic duties. Furthermore, she argued, rather than suffering from the pressures of keeping up with the male students, women were, in fact, victims of the constant reminder that for them education "does not matter."[31]

Clarke's vague recommendation for a special educational program for women was rejected as so impractical that it would lead to little or no education if it were implemented. Mrs. Duffey pointed out that Clarke's plan could only work if each student were subject to a uniform menstrual period. "But each girl has her own time; and if each were excused from attendance and study during this time, there could be neither system nor regularity in the classes." And what of the teacher, who probably also was a woman? "She too requires her regular furlough, and then what are the scholars to do?" Duffey wondered whether Clarke would extend his argument to the home so that wives could leave children uncared for and dinners uncooked for three or four days each month. "I think a concerted action among women in this direction," Duffey wrote, "would bring men who are inclined to agree with the doctor to their senses sooner than anything else."[32]

Most of the critics were unwilling to go as far as a general women's menstrual strike, and simply called for scientific studies to test Clarke's thesis. The first response to their cries came from the Harvard Medical School in 1874, announcing that one of the two topics to be considered for its annual Boylston Medical Prize competition was, "Do women require mental and bodily rest during menstruation and to what extent?" Since the applicants' names were not revealed to the committee, it was possible for a woman to be judged fairly in the competition, and Cambridge friends urged Dr. Mary Putnam Jacobi of New York to apply.[33]

Dr. Jacobi had previously examined the question in "Mental Action and Physical Health," in Anna Brackett's *The Education of American Girls,* one of the four books responding to Clarke in 1874. The article had been written with the general public in mind, and Jacobi had concentrated on causes other than education that might explain female "disabilities," such as "competition, haste, cramming, close confinement, long hours, and unhealthy sedentary habits." For the Boylston competition, which she eventually won, Jacobi sent out 1,000 questionnaires concerning the relationship between rest and general health.[34] She received 286 responses to such questions as how far young women walked; the presence or absence of uterine disease; and the degree and intensity of mental activity in and after school. Tests were also run on a smaller number of women at the New York Infirmary where scientific measurements could be taken on biological responses during the menses and at other times.

Her scientific findings were clearly at odds with the Clarke thesis. More than half, 54 percent, of the women did not experience any menstrual difficulties whatsoever; and, since most of those who did experience difficulties suffered only moderate pain, Jacobi felt there was nothing in the nature of menstruation to imply the necessity, or even

*Figure XVI.    Dr. Mary Putnam Jacobi*
*(Collection of Mary R. Walsh)*

the desirability, of periodic rest for the vast majority of women. In fact, proper physical exercise, along with better nutrition, could do a great deal in Jacobi's view to prevent the development of menstrual pain. Jacobi struck a modern note in her comments when she declared that most women would be better off continuing their normal work patterns rather than taking to their beds to rest. Equally important was her research finding that mental activity was not physically dangerous. In those cases of severe menstrual pain, the women inevitably had some anatomical difficulty causing the disability.

Dr. Jacobi's monograph was followed by a number of studies that appeared to demonstrate clearly that higher education was not injurious to the health of American women. One of the first acts of the Associated Collegiate Alumnae, formed in 1882, was to commission an examination of the health of college women. Questions were drawn up with the help of a group of physicians and mailed to 1290 women college graduates in the United States. The 705 responses were analyzed by the Massachusetts Bureau of Labor Statistics. The Bureau found that 78 percent of the respondents were in excellent health; 5 percent were classified as in fair health; and 17 percent were in poor health. If anything, college women appeared to enjoy better health than the national average, though this is not surprising in view of their economic backgrounds. Similarly, L.H. Marvel, in an article in the journal, *Education,* found that college life, far from being deleterious to woman's physical adjustment, "had resulted in a stronger physique and a more perfect womanhood." In addition, Marvel's statistics demonstrated that the mortality rate for graduates of Mount Holyoke was substantially lower than that for men's schools such as Amherst, Bowdoin, Harvard and Yale.[35]

The debates on women's biology confirm Ruth Hubbard's point that "what questions we ask about the world, in what way and to what end, depends on who is asking, when, and where."[36] The doctors who had a vested interest in proving women's unsuitability for medical careers concentrated their attack on the female reproductive system. Convinced of the correctness of their position, the medical spokesmen repeated each others' prejudices rather than getting themselves involved in any serious scientific investigation of the subject. This course often led to some interesting contradictions. For example, the male medical establishment, though opposing the entry of women into the profession as doctors, welcomed them as nurses. Somehow, the nurses' uniform was a successful antidote to the biological limitations that had been the curse of women doctors. Thus the *Boston Medical and Surgical Journal,* which had objected to women doctors because their delicate health would prevent them from enduring the strain of constant house calls, would later praise the nurses for their constant round of visits to the sick in the slums of the city.[37]

Medical women and their friends, on the other hand, made every effort to support their position through careful investigation. The result was the first series of scientific studies to dispel the myth of female inferiority. But women physicians did more than simply develop arguments in favor of women's entry into medicine. Their research led to a wide ranging investigation of factors related to women's health. I have located 145 scientific articles by women physicians in the period 1872–90 dealing with such subjects as feminine hysteria, hysterectomies, menstrual difficulties, midwifery, and female insanity.

A number of female physicians also played a role in opposing the excessive surgical experimentation on women in the nineteenth century. Dr. Lillian Towslee, in 1903, had charged the medical profession with having gone "mad in the direction of gynecological tinkering, womb prodding and probing." Towslee noted that women were "rarely permitted to have an ache or a pain referable to any other part of the anatomy." Though she was writing long before the recent feminist call for a move away from surgery to solve gynecological difficulties, Towslee concluded her discussion with the same recommendation.[38]

It seems reasonable to assume that if the percentage of women physicians had continued to increase, medical treatment and research on women would have been significantly different. Instead, women experienced a sharp retrenchment in the form of medical school admissions quotas after the mid 1890s. As I have demonstrated elsewhere, this retrenchment preceded the famous Flexner report and appears to have begun just at the time when women physicians were feeling secure as a result of their successful victory over Johns Hopkins.[39] While the retrenchment did not mean that women were entirely shut out of medical schools, they were made to feel unwelcome there and quotas were established, admitting only about five percent women students for almost fifty years. Such small numbers of women doctors have not been able to have a significant impact on medical research or practices. Women's health, therefore, has been almost entirely in the hands (and heads) of male doctors. The few women who entered the profession not only were trained largely by men, but also were forced to "measure up" to the standards laid down by the (male) leaders of the profession in order to survive as doctors.

In 1970, Dr. Frances Norris testified before a Congressional subcommittee that the federal government's failure to investigate sex discrimination in medicine was in large part responsible for the poor quality of medical care given women in this country.[40] Not only has the paucity of women physicians in the American medical community limited the freedom of women to choose physicians of their own sex; it has also determined the priorities of medical research itself. There is a growing awareness of the relationship between the discrimination women face as medical students and physicians, and as patients—both contributing to the present reliance on radical mastectomies to treat breast cancer; the excessive frequency of hysterectomies; the lack of concern for the hazards of the various methods of birth control and of hormone treatments; and the generally deficient health care for women.

The  dramatic  increase  in  the  number  of  women  entering  medical school  during  the  past  few  years,  up  more  than  700  percent  since  1960, marks  the  first  real  progress  for  medical  women  since  the  late  nineteenth century.  It  would,  of  course,  be  unfair  to  expect  the  new  crop  of  medical women  to  transform  the  whole  of  American  medicine.  However,  the pioneering  work  of  their  sisters  almost  a  century  ago  provides  a  model for  what  can  be  accomplished  by  research  done  from  a  woman's  point of  view  and  focused  on  issues  of  concern  to  women.  Yet  the  failure  of nineteenth-century  women  physicians  to  sustain  their  progress  serves as  a  warning  to  those  who  argue  that  the  battle  of  women  in  medicine has  now  been  won.

## Notes

1. These and other statistics cited later in this essay are based on handcounts of female names on lists from medical colleges which were broken down by gender. Mary Roth Walsh, *Doctors Wanted: No Women Need Apply* (New Haven: Yale University Press, 1977) contains detailed tables on women physicians and medical students on pp. 185, 186 and 193.
2. Newspaper dates are 1894, 1886 and 1881, respectively.
3. For a detailed discussion of these novels, *see* Mary Roth Walsh, "Images of Women Doctors in Popular Fiction: A Comparison of the 19th and 20th Centuries," *Journal of American Culture,* 1:2 (1978), pp. 276–284.
4. Mary Roth Walsh, "Feminism: A Support System for Women Physicians," deals with this phenomenon in detail. *See* the *Journal of the American Medical Women's Association,* 31 (1976): pp. 247–250; also, Chapter 3 of *Doctors Wanted . . . ,* "A Feminist Showplace."
5. Mary E. Putnam Jacobi, M.D., "Social Aspects of the Readmission of Women into the Medical Profession" *Papers and Letters Presented at the First Women's Congress of the Association for the Advancement of Women, October, 1873* (New York: Association for the Advancement of Women, 1874), p. 177.
6. Marion Hovey to Harvard Medical School, March 21, 1878, Harvard Medical School Dean's Records; Agnes Vietor, *A Woman's Quest: The Life of Marie Zakrzewska,* (New York: D. Appleton and Co., 1924); *Doctors Wanted . . . ,* p. 173.
7. *Nation,* (12 February 1891), p. 131; *Century* (May 1891), pp. 632–36.
8. A bibliography of major publications on this topic is in *Doctors Wanted . . .* pp. 4–8; fn. 7–18. *See also* Jean Donnison, *Midwives and Medical Men: A History of Inter-Professional Rivalries and Women's Rights* (New York: Schocken Books, 1977). The most recent interpretation of the reasons for American women's loss of dominance in midwifery is Datha Clapper Brack's essay in this collection.
9. *Boston Medical and Surgical Journal,* 75 (1866), pp. 191–92.
10. C. D. Meigs, "Lecture on Some of the Distinctive Characteristics of the Female, delivered before the class of the Jefferson Medical College, January 1847" (Philadelphia, 1847), p. 67.
11. *Annual Reports of the New England Hospital* (1866), pp. 10–11.
12. For more detailed information on Storer's life history, see the references in *Doctors Wanted . . .* pp. 113–115, fn. 16 and 17.
13. Malcolm Storer, "The Teaching of Obstetrics and Gynecology at

Harvard," *Harvard Medical Alumni Association,* 9 (1903), pp. 439–40.

14. *See* circular, dated 1864 with letter from John H. Stephenson endorsed by Drs. Horatio Storer, Walter Channing, C. P. Putnam, S. Cabot, and Henry I. Bowditch in New England Hospital Collection, Schlesinger Library, Radcliffe College.

15. Frederick C. Irving, *Safe Deliverance* (Boston: Houghton-Mifflin Co., 1942), pp. 114–116.

16. Vietor, *A Woman's Quest . . .* , p. 226.

17. G. J. Barker-Benfield, *The Horrors of the Half-Known Life* (New York: Harper and Row, 1976), pp. 91–119; and Seale Harris with F. H. Brown, *Woman's Surgeon* (New York: Macmillan, 1950), pp. 235–272.

18. Horatio R. Storer, "Report of the First Regular Meeting of the Gynecological Society of Boston, January 22, 1869," *Journal of the Gynecological Society of Boston* (July 1869), p. 14.

19. *Boston Medical Surgical Journal,* 81 (1869), p. 345; Hiram Corson, *Brief History of Proceedings in the Medical Society of Pennsylvania to Procure Recognition of Women Physicians* (Norristown, 1894); *Evening Bulletin* (Philadelphia, November 8, 1869); "Women as Physicians," *Philadelphia Medical and Surgical Reporter* (April 1867), p. 2; *The Press* (Philadelphia, November 18, 1869); Clara Marshall, *The Woman's Medical College of Pennsylvania* (Philadelphia, 1897), pp. 17ff.

20. *Boston Medical and Surgical Journal,* 81 (1869), p. 345.

21. *Ibid.,* pp. 345–346.

22. *Ibid.,* p. 346.

23. Edward H. Clarke, *Sex in Education; or, A Fair Chance for the Girls* (Boston: James R. Osgood and Co., 1873); M. Carey Thomas, "Present Tendencies in Women's College and University Education," *Educational Review,* 25 (1908), p. 68; Dorothy Gies McGuigan, *A Dangerous Experiment: 100 Years of Women at the University of Michigan* (Ann Arbor: Center for Continuing Education of Women, 1970), p. 38; Lilian Welsh, M.D., *Reminiscences of Thirty Years in Baltimore* (Baltimore, 1925).

24. *Sex in Education,* pp. 81–82.

25. *Ibid.,* pp. 23, 41.

26. *Ibid.,* p. 63.

27. *See* Oscar Handlin, "The Horror," in *Race and Nationality in American Life* (New York: Doubleday Anchor Books, 1957), pp. 111–132.

28. *Sex in Education,* p. 140.

29. Eliza Bisbee Duffey, *No Sex in Education: or, an equal chance for both girls and boys* (Syracuse, 1874), p. 117.

30. Julia Ward Howe (ed.), *Sex and Education. A Reply to Dr. Clarke's "Sex in Education"* (Boston, 1874), pp. 191–92.

31. *Ibid.*, pp. 27–28.
32. Duffey, *No Sex in Education* . . . , pp. 115–16, 97.
33. Alice C. Baker to Mary Putnam Jacobi, Cambridge, Massachusetts, 7 November 1874, folder 18, Jacobi papers, Schlesinger Library, Radcliffe College.
34. This survey was published in book form: Mary Putnam Jacobi, *The Question of Rest for Women During Menstruation* (New York: G. P. Putnam, 1877).
35. Louis H. Marvel, "Why Does College Life Affect the Health of Women?" *Education*, 3 (1883), p. 501.
36. Ruth Hubbard, "When Women Fill Men's Roles . . .," *Trends in Biochemical Sciences*, 1 (1976), pp. N 52–53.
37. *Boston Medical and Surgical Journal*, 76 (1867), p. 217. *See* Walsh, *Doctors Wanted* . . . , Chapter 4, for a fuller discussion of this whole issue.
38. Lillian G. Towslee, M.D. *Women's Medical Journal*, 13 (1903), p. 121.
39. A detailed discussion of this retrenchment process appears in *Doctors Wanted* . . . , pp. 186–206.
40. *Hearings before the Special Subcommittee on Education of the Committee on Education and Labor, House of Representatives, 91st Congress, 2nd Session on Section 805 of H.R. 16098, Part 1* (Washington, 1970), p. 511.

[Mary Roth Walsh *is Associate Professor of Psychology at the University of Lowell. In 1977 she published* Doctors Wanted: No Women Need Apply *with Yale University Press, a book that describes and analyzes barriers to female achievement in medicine over the years 1835–1975. She has actively worked on the problems of women in medicine since 1973 when she became research manager at Harvard Medical School for the Committee on the Status of Women. Since 1977, she has led leadership training sessions for women physicians, lectured nationally on their behalf, and has become the Director of Academic Curriculum for a Women's Educational Equity Act project to help women physicians sustain the momentum of their current success. She is also co-director of a shelter for Battered Women in the City of Lowell, Massachusetts.*]

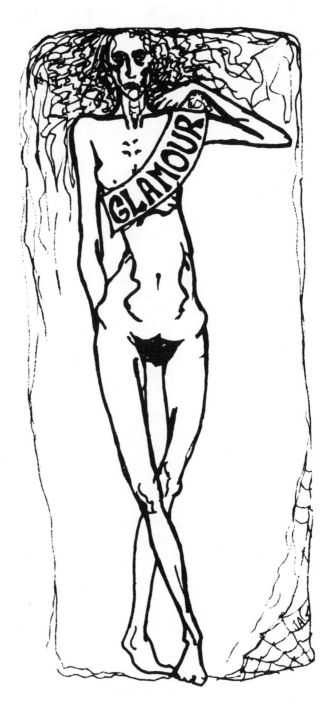

Judy Greenberg

# Vicki Druss and
# Mary Sue Henifin

---

# Why Are So Many Anorexics Women?

*if she could take up less space
she would*

*if she could eat herself
to the white precision
of a fishbone on a plate
she would do it*

*if she could pick herself clean
if she could suck herself thin
as straw   if someone would take her
and spin her to straw again*

*in lumpy arms and breasts
she who was once all spine
is a rootcellar gone haywire
bruised rutabaga  veiny cabbage*

*she feels huge to herself
if terror could lace her tighter
she knows they will love her more
for less and less*

*if absence makes them grow fonder
she grows more absent*

*if being human is
mash in the mouth   gristle and spit
the red hand stuffing the prongs
in the red hash faces*

*she wants no part of it*

*she wants cadaverous wings
and hips like blades*

*she will be silvery*
*as mercury under glass*

*she will dance in the mirror*
*like a transparent*
*sheath for her knives*

*she will say*
*there is no pain*

*she will never*
*feel hungry*

>                    —Monica Raymond,
>                        "Anorexia"

It is the first of the month, and on thousands of newsstands across the country, we see the new month's wares. In these enlightened times, some women show signs of resentment as they pass by, casually glancing at the covers of popular "women's" magazines that line the shelves. Yet even many women who understand the damage done us by articles like "Do Your Hair and Make-up Like Our Cover-girl and Dazzle Him," still linger hopefully over "Complete Diet Guide For People On the Go" and "13 Marvelous Diets That Really Work."

It is the first of the month, and millions of women across the country are being fooled into believing that they have weight control problems to which answers are available for 50 cents at any newsstand. A great deal of time and *man*power have been invested to convince women that we have eating disorders whose solutions have finally been discovered. While we have become generally more aware and wary of the "suggestions" offered by the media concerning our life-styles, the advertisements aimed at women with eating control problems have, if anything, grown more influential. They are effective because they address an area which we have long been pressed not to explore deeply. It is difficult to stop hoping that "maybe this time it will be different, maybe this time one of the 13 marvelous diets will actually work." Yet it is essential that we give up wishing, and instead begin to ask why women place so much emphasis on body weight.

Let us be clear: the usual weight loss techniques are harmful, no matter what the outcome. If a woman tries to implement one of the "easy" methods and fails, every effort is made to assure her that it is *her* fault—not the method's. And if she succeeds, her "success" story, no matter how short-lived, is publicized not only to obtain more customers, but also to serve as a deterrent to her further understanding of whether she, in fact, *has a* "weight problem," and if so, why. Only if we address the social context surrounding the issue of our "weight anxiety," instead of manipulating our bodies to conform to the parameters of newsstand beauty, can we hope for a lasting solution to the eating disorders that afflict some women.

Although most popular attention has been directed at "overeating" and "overweight," it is useful to look at the other extreme, self-starvation or *anorexia nervosa.* We will then gain some insight into what may motivate some women to use extreme forms of "body language." Anorexia has been described as "self-inflicted starvation in the absence of recognizable organic disease and in the midst of ample food."[1] It occurs mostly in young, upper-middle-class white women (for whom thinness is a prerequisite for social status), and appears to be on the rise in extremely weight-conscious societies like our own. It is usually observed in girls or young women between the ages of 11 and 35, most commonly between the ages of 15 and 23.[2] Approximately 90 per cent of all anorexics are female.[3]

Anorexia has been classified in medicine as a "functional" disorder, which means that no structural or organic change has been found to precipitate it. The excessive weight loss is thought to result from an active refusal to eat. A criterion that has been cited as "sufficient" for a diagnosis of anorexia is the loss of 10 to 20 per cent of the "ideal" body weight. In extreme cases, adults have shrunk to 41 pounds. The "average female anorexic patient, according to published medical records, weighs 122 pounds prior to the beginning of self-starvation and drops to 78 pounds, a loss of 44 pounds in the course of the 'illness.' "[4]

Anorexics generally lose weight through a deliberate reduction of calories, although they rarely stop eating completely. Many anorexics "supplement" their dieting with laxatives and forced vomiting. Some observers have suggested that the self-induced vomiting expresses the ambivalence anorexics have about food and eating, since vomiting occurs after an anorexic has given in to her intense desire to eat, following days or weeks of insufficient food intake. After the "food binge," she may atone for her "sin" by giving up the food she just ate to avoid the terrifying alternative of gaining weight.

Amenorrhea is another criterion often used to help diagnose anorexia. Yet, according to Dally, while 38 per cent of his cases stopped menstruating after they started losing weight, the majority had stopped menstruating *before* they began dieting.[5] These statistics suggest that high levels of emotional conflict and stress may be present before anorexics initiate their regimens of self-starvation.

In a study of thirty anorexic patients, the opinion was tabulated of each patient's psychiatrist on the sources of emotional conflict that may have motivated the onset of anorexia (more than one source of emotional conflict could be attributed to a patient). The "reasons" given by the psychiatrists (in typical psychiatric jargon) are presented in Table I.[6]

*Table I*

| REASON | No. of Psychiatrists Who Give This Reason |
|---|---|
| Means of winning familial attention and privilege | 14 |
| Indication of fixation at or regression to the oral level | 13 |
| Angry gesture toward parental figures | 11 |
| Defense against incestuous wishes for father | 10 |
| Way to remain physically a child | 10 |
| Means of expiating guilt for anger at parental figure and other means of controlling parents, siblings and others | 7 |
| Defense against oral impregnation fantasies and unconscious pregnancy wish | 6 |
| Defense against "oral aggression" | 5 |
| Result of retroflexed rage at parental neglect leading to depression and self-destructive tendencies | 3 |
| No comment by therapist | 4 |

In addition to being wary of the "causes" cited in this survey, which are, of course, only guesswork, one questions whether they are causes at all, rather than by-products or effects. For example, does "means of winning familial attention and privilege," or "angry gesture toward parental figures" point to a *source* of emotional conflict for anorexic individuals, or does it simply describe an anorexic's response to a deeper source of conflict. An answer to *why* large numbers of women, in particular, resort to body manipulation in order to win familial attention might explain a great deal more.

Like the newsstand diet pushers, these psychiatrists fail to see eating disorders as symptoms of perhaps wider social, rather than individual, problems. Of all the "explanations" on the list, only one—"defense against oral impregnation fantasies and unconscious pregnancy wish"—addresses even indirectly the fact that most anorexics are *female*. This fact, so little discussed in the literature on anorexia, raises some interesting and crucial questions.

For instance, why is it that only women tend to manipulate the size and shape of their bodies as an expression of their emotional states? Although some men may also employ this means of communication, what other expressive channel is there to which males have access, and which they prefer? What types of conditioning patterns create differences in the ways males and females regard and use their bodies?

It is significant that in anorexic women the much admired "feminine curves" are obliterated. One opinion that is therefore prevalent among researchers is that anorexia is "a way to remain physically a child."[7] *Why,* one is prompted to ask, should so many more women than men in our society *want,* physically, to remain children? Does this suggest that some women so dislike being regarded, first and foremost, as sexual creatures that they prefer to regain a "child-like" shape? If so, this is not a personal neurosis, but reflects an accurate perception of the realities of many women's lives in our society.

And what can we make of "defense against incestuous *wishes* for father, and unconscious pregnancy *wishes*" (emphasis ours) being put forward as "reasons"? It seems just another case of "blaming the victim," typical of the kind of psychologizing that translates a woman's *fear of rape* into a fear of the desire to be raped. Perhaps more attention should be given to the fact that 23 per cent of the anorexics in this study admitted to having been subjected to seductive behavior by their fathers, including "fondling, kissing, and deliberate fondling of the breasts and genitals."[8]

These are only a few of the issues that need to be examined regarding the origins of anorexia. That they haven't been addressed, that most research on anorexia has tended to ignore or gloss over the fact that it is mostly a female disease, should not surprise us. Psychologists traditionally have been loathe to regard female "personality disorders" as anything broader than the malfunctions of individuals in a well ordered society. Until prodded by the recent rebirth of feminism, they have tended not to consider such "disorders" as perhaps appropriate responses to women's objective situation in a disordered society.

## Body/Mind/Society

For many women in our culture, their body image often *is* their definition. This is not the distortion of a sick mind; it is literally true.

And the media constantly remind us: it doesn't matter who you are or what you achieve—you must *look* "right." Some women can reject these definitions, but most of us internalize them. Therefore, for many of us, it is not only the society that defines us by our bodies, but also we who come to define ourselves that way.

Stop and think: how did you feel when your body first became "womanly"? When your mother/father/brother first noticed the change? When the first man whistled at you, or made a suggestive remark? Were you proud of your body then? Did you feel full of confidence that its changes were opening for you doors to a happy, adult life?

These are the questions we must ask ourselves and each other. And when we find our own, honest answers we may also find the roads back to our selves and to our bodies. Large or small, fat or thin, we must, first of all, accept our bodies as our own. Then, if we wish, we can begin to think about whether we need, or want, or are able to change them.

## Notes

1. Ernst Joachim Meyer and F. Feldman, *Anorexia Nervosa Symposium,* (Stuttgart: G. Thieme, 1965), p. 71.
2. Peter Dally, *Anorexia Nervosa,* (New York: Grune and Stratton, 1969), p. 6.
3. Eugene L. Bliss and C. H. Hardin Branch, *Anorexia Nervosa* (New York: Hoeber, 1960), p. 6.
4. *Ibid.,* p. 27.
5. Peter Dally, *Anorexia Nervosa,* p. 6.
6. Christopher V. Rowland, ed., *Anorexia and Obesity* (Boston: Little Brown and Co., 1970), p. 70.
7. One therapist wrote about the anorexics he had studied, "these people were afraid of adult, genital sexuality and could avoid menses, breast and hip development . . . and general physical attractiveness by starving themselves." Another pointed out that his anorexic patients were "in tremendous conflict over whether to be 'a big girl or a little girl' in both a specific and general sense." A prepubescent anorexic girl was described as not wanting "to dress 'like grown up girls.' " (Christopher V. Rowland, *Anorexia,* p. 72).
8. One physician, Dr. William Langford of Babies Hospital in New York City, remarks that in the 13 cases of anorexia he had treated, the father-daughter relationships had a "striking degree of necking and petting in them." He then goes on to note that both parent and child were extremely reluctant to discuss this and concludes by

saying that no therapy work was done on this issue! (Christopher V. Rowland, *Anorexia*, pp. 115, 116.)

[Vicki Druss *is a medical student on leave from the Medical College of Virginia. She is currently studying and working in Israel. She hopes eventually to understand more about women and eating disorders and welcomes fellow searchers along that path.*]

[Mary Sue Henifin *has studied plant ecology and has helped teach courses in botany and on biology and social issues at Harvard. She was active in the collective that started the Radcliffe-Harvard Women's Center, where she taught a course on feminist theory. She has co-produced "Move Sisters," a feminist radio show. She is co-editor of a book on conservation of rare plant species and has written several papers on this subject. Her interests include history of women in science, women's use of plants and cultural evolution, and ecology and feminism. She thanks her friends in Oregon and Cambridge (particularly the women basketball players, cosmic maintenance persons, and women's history enthusiasts) for their support.*]

# Emily E. Culpepper

## Exploring Menstrual Attitudes

*My Friend*
*Sickness*
*Have Your Period*
*Those Days*
*On The Rag*
*Menstruation*
*The Curse*
*Got Your Period*
*Fall Off The Roof*
*Female Ailment*
*That Time Of The Month*
*Aunt Martha*
*Red Tide*
*Flow*
*Monthlies*
*Menses*
*The Duchess*
*Sweet Secret*
*Red Moon*

*I am older than this age*
*The moon calls forth my blood*
*    even as it rules the ocean tide*
*I dream of ancient women*
*    who did not apologize*
*For their moon stains*
*Or their Way of Living*
*I pray their like will come on earth again.*

—Lanayre Liggera, "Invocation"

## Introduction

Menstruation is a normal, frequent occurrence for nearly one half of most women's lives, and menopause marks its equally normal cessation. However, due to women's devalued position throughout patriarchal history, we have little knowledge about these experiences.[1] Beginning with some of the oldest cultures and continuing into our own time, menstruation has been the subject of cultural and religious taboos.

Ancient, patriarchal societies often isolated women during menstruation and labeled them "unclean," even evil and dangerous. Frequently the menstruating woman was confined to a special hut. Often she was prohibited from participating in the religion that was usually at the center of tribal life. The many complex regulations of her behavior while menstruating were difficult to follow (e.g., not to touch food, not to look at anyone), and included penalties if she failed to observe them properly.

Such customs and beliefs arose as part of the deep fear of women inherent in, and perpetuated by, male-dominated societies that considered the male body the norm. Women have been judged as different and alien—the inferior Other. Women's sexual organs and biological functions have been viewed as disgusting and dirty. Many women have internalized such attitudes and remain ignorant and fearful about their bodies and especially about what healthy, normal menstruation is or could be.

The belief that menstruation is a female "illness" ("ailment," "sickness," "weakness," "curse") is still widespread. Even when not explicit the embarrassment and uneasiness that surround menstruation and menopause convey a negative connotation. Many girls reach their first period with little or no preparation; many women approach menopause with anxiety bred of ignorance and rumor.

Many women do experience uncomfortable sensations (such as cramps or moodiness) with menstruation. Yet, with medicine and health care largely in the hands of male doctors (most of whom bring to their work erroneous preconceptions about women), there continues to be little understanding of the sensations that are part of the menstrual cycle. Medical theories are contradictory, ranging from the view that all cramps are emotional or psychological, to the view that most cramps have direct physical causes. The theories are not substantiated and, by and large, have not led to successful treatments for menstrual discomfort. Moreover, there is little or no acknowledgment of the positive sensations (e.g., creative surges and erotic energy) that many women experience with menstruation. There is also widespread belief that the menstrual cycle makes women more emotional than men, although this is unsubstantiated. Some studies[2] suggest that when such behavior occurs, it is an unconsciously learned, conditioned response. However, other studies[3] indicate that males, too, experience (but are taught to deny) cyclic rhythms similar to those that women are encouraged to exaggerate and identify with their menstrual cycle.

Clearly, women must begin to define our own experiences and search for our own answers. We need to keep our own charts and records of our menstrual cycles to discover what is healthy and what is unusual for each of us. We need to explore current attitudes about menstruation (our own and the attitudes of those around us) and see how these need to change.

Since menstruation and menopause are nearly universal female experiences, the search for a new approach will begin by exploring different directions. I believe that a common starting point lies in women's free inquiry while sharing information with each other. In my own studies of menstruation, I am uncovering a vast, unacknowledged tradition of women's stories and folklore about menstruation. We need to share our experiences as the first step in a woman-identified exploration of this aspect of ourselves. After all, each of us menstruates about 1500-1800 days (or 4 to 5 years) of our lives!

One way to open up space for new considerations of our menstrual experience is to highlight different attitudes toward menstruation. First, as background, let us look at ancient Hebrew teachings about menstruation. These profoundly influence much of western culture and their negative content is still with us. I do not intend to imply that the Hebrew attitudes were worse than those of many other ancient cultures. Unfortunately, one could compile many such studies, documenting the diversity of perverse cultural and religious hostility toward women and menstruation.[4]

Next is a fantasy story—a social/science fiction visit to a future time in which menstruation has high social value as a special form of consciousness and in which there are celebrations surrounding menstruation and menopause. Many kinds of writing and research about menstruation are needed and this frankly experimental story will appeal to some and not to others. My purpose here is to evoke and encourage an uninhibited, wide range of female thinking and exploring. With our long history of restrictive and repressive traditions, fantasies such as this one can help open new doors of perception and new ways of experiencing ourselves. Some women are working toward just this kind of future, while others are developing the controversial technique of menstrual extraction. My own personal/political point of view is to stress at *this* time the crucial importance of investigating multiple, sometimes conflicting, directions.

We must find ways to respect and encourage (even, and indeed *especially,* with serious disagreements) each other's best efforts to create, imagine, invent, explore, name, and define our own menstrual experiences.

We do not need to create a new "feminist mystique" of menstruation. Rather, let us develop woman-identified approaches to the experience and study of menstruation, which after all involves the study of our Selves. To this end, there is attached an annotated selection of sources for readers who want to search further. We will break out of the ignorance and confusion veiling menstruation only by collecting our own data, sharing our many stories, developing our own health lore, and transforming our stereotyped attitudes and ideas. The implications of this menstrual exploration are simultaneously biological, spiritual, emotional, philosophical, and political. We must find our own way of doing it, but, together, let us find out about ourselves.

## Niddah:  *Unclean or Sacred Sign????*[5]
### *Ancient Hebrew Teachings on Menstruation*

Information about ancient Hebrew attitudes toward menstruation comes from three main sources: Hebrew scriptures (*Torah* and *Mishnah*), focusing on Leviticus 15:19–24, 18:19, and 20:18; commentaries on the *Torah* and *Mishnah* passages; and other discussions about Hebrew scripture and custom. Because of their central significance, some of the Leviticus passages are reproduced here.

*Leviticus 15:19–24*

And if a woman have an issue, and her issue in her flesh be blood, she shall be in her impurity seven days: and whosoever toucheth her shall be unclean until the even.

And everything that she lieth upon in her impurity shall be unclean; every thing also that she sitteth upon shall be unclean.

And whosoever toucheth her bed shall  wash his garments, and bathe himself in water, and be unclean until the even.

And whosoever toucheth any thing that she sitteth upon shall wash his garments, and bathe himself in water, and be unclean until the even.

And if he be on the bed, or on anything whereon she sitteth, when he toucheth it, he shall be unclean until the even.

And if any man lie with her, and her impurity be upon him, he shall be unclean seven days; and every bed whereon he lieth shall be unclean.

*Leviticus 18:19*

And thou shalt not approach unto a woman to uncover her nakedness, as long as she is impure by her uncleanness.

*Leviticus 20:18*

And if a man shall lie with a woman having her sickness, and shall uncover her nakedness—he hath made naked her fountain, and she hath uncovered the fountain of her blood—both of them shall be cut off from among their people.[6]

When you're wearing a tampon you don't worry about odor. But should you?

Primarily, menstruation is BLOOD,* and this fact makes it highly charged in religious significance. Blood in all its manifestations is a source of *mana*—that is, it represents the power of elemental, natural forces—inspiring awe and fear. Blood sacrifices are some of the earliest known

---

*Literally the menses consists of more than blood, but blood colors the whole flow red. Thus, people often regard it and react to it as blood.

religious rites, spanning many diverse cultures. In Levitical times, blood was associated with the mysteries of life and death, and the Hebrews believed that the very essence of life was in this bright crimson fluid. For this reason, blood was used in their atonement rituals. The Hebrews are commanded not to eat blood, "for the life of the flesh is in the blood,"[7] but rather to reverence it. Animal blood was applied to the altar and to the symbols of God's presence. Meat had to be carefully drained of blood before being eaten.

In this context then, how is menstrual blood regarded? This blood has the added aura of reproductive and sexual powers. The creation of new human beings was associated with it, and it was considered the "stuff" from which a child was made. It issues from the vagina and therefore was associated with *female* sexuality. While the primary male experiences of bleeding were through illness, injury, war, and death, women could bleed regularly and not die from it.

However, although menstruation is a normal function of the female body, it was looked upon as a sickness (Lev. 20:18), with abnormal menstrual flow (real sickness) defined as a worse sickness (Lev. 15:25–31). In the early literature, sickness does not seem to mean cramps, or to imply that the menstruant behaved differently. (This seems to be a later historical development, and an explanation for menstrual taboos retroactively applied by some rabbinical commentaries.) Rather, sickness seems to mean that a menstruant simply *is* in another state of being. Perhaps the idea of sickness also reflects ancient myths that the vulva and/or menstruation are the results of a wound.

Menstrual blood, then, is unique, and therefore its characteristics are carefully discussed.* Also important to the Hebrews was the fact that menstruation is "an issue" (a "flow" or "flux") and all "issues" or excretions from the human body (female and male) were *Unclean*. We must look at what *Uncleanness* meant to understand the significance of naming menstruation and the menstruant herself *Unclean*.

There were many causes of Uncleanness and "the distinction between clean and Unclean is to be found as far back as we are able to trace the history of Israel."[8] What Uncleanness meant to the Hebrew people was intimately bound up with their idea of the *Holy*. Both the Unclean and the Holy were taboo; that is, they were both Sources of power to be feared, respected, and set apart because their power was infectious.[9] "The uncleanness of girls at puberty and the sanctity of holy men . . . are only different manifestations of the same mysterious energy."[10]

---

*The *Niddah* (a section of the Sedar Teharoth of the *Mishnah*) has a detailed discussion of the various shades of colors of menstrual fluid.

Eventually some Sources of power came to be seen as positive and Holy, others as negative and Unclean; but traces of their common origin are still apparent in the Pentateuch, which sets aside elaborate rituals to deal with both.[11]

There are four main types of Uncleanness in the *Torah*: female and male functions of reproduction, foods, leprosy, and corpses. Menstruation was one of the most serious Uncleannesses, and it was used in the Pentateuch as an analogy (an important principle of interpretation) to show how seriously Unclean other things were. In some instances, the menstruous woman herself was used as a metaphor for horrible filth, e.g., Isaiah 30:22, "cast them [idols] away as a menstruous

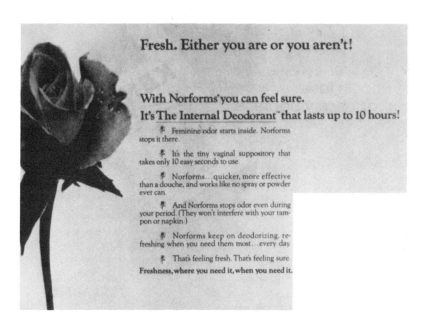

**Fresh. Either you are or you aren't!**

**With Norforms you can feel sure.
It's The Internal Deodorant that lasts up to 10 hours!**

★ Feminine odor starts inside. Norforms stops it there.

★ It's the tiny vaginal suppository that takes only 10 easy seconds to use.

★ Norforms... quicker, more effective than a douche, and works like no spray or powder ever can.

★ And Norforms stops odor even during your period. (They won't interfere with your tampon or napkin.)

★ Norforms keep on deodorizing, refreshing when you need them most... every day.

★ That's feeling fresh. That's feeling sure.

**Freshness, where you need it, when you need it.**

woman." An entire section (Masechtoth) of the *Mishnah,* named "The Menstruant" or "Niddah," is primarily concerned with menstruation. A woman had to take great care to know when she was menstruating and to observe all the laws about menstruation, lest she transmit her Uncleanness to others, for not only was her menstrual blood Unclean, but *she herself* was Unclean. She was a *primary* Source of Uncleanness— not only did what she touched become Unclean, but that object *itself* would then transmit Uncleanness.

A menstruant transmitted Uncleanness "Through six removes" (i.e., six touches distant from original contact with her). Comparison to other

Uncleannesses is helpful to emphasize the power of *Niddah*. Men with abnormal discharges (i.e., probably gonorrhea), women with abnormal discharges, women after childbirth (twice as long for a female child), lepers, human corpses, dead reptiles, animal carrion, and incestuous sexual relations: all these were discussed in the *Mishnah* and the commentaries as being Unclean *like the menstruant was unclean*.[12] Menstrual blood was Unclean wet or dry (as were the remains of a corpse), but

semen was only Unclean when wet.[13] Although semen is the normal, male body-flow most frequently compared to female menstrual flow, it was definitely *not* as Unclean and conveyed Uncleanness only to four removes. The commentary of Maimonides stated that while a man conveyed Uncleanness through semen and a woman through blood, these cannot be used analogously in interpreting the Law[14] because "the flux of a man is different from that of a woman."[15]

The most severe rule of all was the repeated prohibition against intercourse during a woman's Unclean time. Leviticus 15:24 classified this as rendering one *ritually* Unclean (which made one liable for a sin offering). Leviticus 18:19 and 20:18, however, which are part of the older "Holiness Code," prohibited coitus during menstruation as *morally* unclean, and prescribed the death penalty for *intentionally* breaking this law. The language of Leviticus 20:18 is particularly powerful, describing the forbidden coitus during menstruation as "uncovering the fountain of her blood." This act not only brought one in contact with Uncleanness, it deliberately associated one with the very Source-in-itself of taboo, Unclean life-blood.

Ezekiel 18:6 lists observance of this Law as one of those required of the just man. The distinction between willful and unintentional transgression was apparently quite strict. A great many verses in the *Niddah* and other sections of the Teheroth were concerned with detecting menstrual blood right away, this being the woman's responsibility so that such a grave transgression would not occur through laxity. She was required to examine herself twice a day (morning and evening) and

before intercourse with "test-rags" to be sure she was clean from any sign of blood. Women of a priest's household were required to examine themselves before eating the Heave Offering. There were also references to males using a test-rag, apparently after intercourse. If a test-rag showed blood at any time, some conservative sources (i.e., Hillel) say a woman was retroactively Unclean back to the time of her last clean examination, thus rendering any intercourse since *then* ritually Unclean. If a woman and man found blood immediately *after* intercourse, they were *both* ritually unclean. A man who had (unintentional) intercourse with a menstruant conveyed Uncleanness as much as when he had an abnormal discharge. A woman and man who had (unintentional) intercourse during menstruation both had to be immersed in a ritual bath (mikvoth), as did everything they touched. These laws about the menses were highly specific, and implied that intercourse during the menses, occurring through negligence in observing precautions, was as serious an offense as intentional coitus during the menses.

The degree of a menstruant's Uncleanness seems to have increased historically. The period of a woman's impurity in the Pentateuch lasted for seven days, counting from the first day of the flow. By the time of the *Mishnah,* this became twelve days. Still later, as an "added margin of pious carefulness," a husband's separation from a wife for twelve hours was observed, prior to the expected onset of the menses (samukh l'vosteh). A bath of ritual purification (mikvath) for the menstruant was not clearly required in the Pentateuch, but became clearly required in the *Mishnah.*

There were other indications of the strong significance attached to the menstruant. One section (the Moed Shabbeth) of the *Mishnah,* said that ignoring the laws of the menstruant was one of three reasons women died in childbirth! In the Nesikin, Aboth urges for study the rules about the onset of menstruation because of its central importance and complexity.

The word "niddah" itself attests to the importance of its subject. The Torah was euphemistic in referring to sexual relations and genitals. (There is in fact no word for the female genitals and a word for penis occurs only twice.[16]) In spite of this, the (negative) religious and ritual importance of the menstruant and the rules about her were so great that the term "niddah" was coined and used frequently. Blackman speculates that the word comes either from a verb meaning "to flee," or from a verb meaning "to banish." He lists its meanings as "isolation, removal, separation, state of uncleanness, menstrual uncleanness, menstruation; a menstruant."[17] Other sources agree on this meaning of "separation," which is sometimes translated as "shunning."

Custom also attests to the importance of menstruation and the power of the prohibitions surrounding it. Orthodox Jews today are expected to observe the "niddah" laws (the 12-day version).[18] Popular Jewish folklore perpetuates the belief that coitus during the menses is unhealthy for males. Young Jewish women report that, often, even nonreligious Jewish males show a marked aversion to menstruation and to intercourse during menstruation.[19] It is a fairly prevalent custom for a Jewish girl to be slapped by her mother at her first menstruation. Reports of the custom range from an unexplained hard slap to a loving pat on the cheek. It seems to be an unofficial, ritual way of marking the significance of menstruation—an ambivalent significance. The Hebrew attitude that menstruation is basically a negative, unclean occurrence still survives.

# My husband won't live with me five days each month

Why wear more than you have to?

What then did this highly charged Uncleanness mean for Hebrew women? It would seem to have been a great burden on them to observe all the *Niddah* laws, with the restrictions these placed on their independence and participation in the life of the community. *Uncleanness isolated one from God.* This was clear in the Pentateuch and the *Mishnah.* God was the Most Holy One. Not only did Uncleanness disqualify someone from religious worship, but "abstinence from uncleanness is part of the way of holiness."[20] Maimonides related bodily cleanliness to sanctity of the soul. It was in this value-laden context that Hebrew women lived about one-fourth of their adult lives (closer to one-half for Orthodox women observing twelve-day "niddah") in a condition named Unclean. Women were, in themselves, regarded as a Source of Uncleanness. Clearly, this isolated Hebrew women more frequently than men from others and from the God they worshipped; and this for a natural function. Indeed, with Yahweh referred to as male and as a father, this constituted a *primordial* separation and alienation for women who existed in a subordinate tension with the Source of Holiness. Orthodox rabbis have claimed that "niddah" laws were a way of showing respect for women, a "hygenic necessity," and a result of a so-called natural repugnance! All of the later historical justifications, however, seem to me insufficient, for they fail to reckon with menstruation as a feared Source of power.

My basic hypothesis, then, is that the intense Hebraic negation of menstruation reveals a fear of it as a Source of power in women, and suggests the possibility that menstruation was formerly a Sacred Sign. Clearly, the Pentateuch's language about menstruation reveals the awe of its male authors at the phenomenon of the regular, cyclic, healthy issue of blood from females. It was a primary body-experience that men did not have, yet it manifested attributes of the Holy One whom they were striving to obey and worship. They conceived their God as a Source of life and the life-power of blood figured prominently in their religious understanding. Thus the Levitical expression about menstruation, "fountain of blood," has strong overtones of meaning "Source of life." The separation of menstruating women and the prohibition and fear of contact with them (especially the intimate, vulnerable, *mana*-charged contact of sexual intercourse) reveal fear of a power that is not understood, but cannot be ignored. Menstruation was a body-sign that women possessed and that was simultaneously associated with the powers of blood, reproduction, sexuality, and cyclic, recurring time.

Significantly, of the four aspects of this body-sign, the Hebrew literature deals with the first three, but does not deal with the fourth, the recurring, cyclic nature of menstruation. This silence is astounding in an historical time abounding with other religions that held cyclic concepts of life, time, sacredness, and reality. Women with their magical menses visibly participated in the rhythm-powers of the cosmos to which men are subject. Such is the framework within which the Hebrew community lived and within which *males* were trying to comprehend reality. The fact that women had a primary body-event around which one could orient time seems to have been too threatening for Hebrew males to accept as a positive element in their universe.

Perhaps it is in this context that we can understand circumcision as an intentionally created male body-sign. Like menstruation, it is associated with reproduction, sexuality, and also somewhat with blood—

since it involves cutting on the body, although it is not a naturally occurring flow of blood. Rather than being a cyclic, recurring sign, however, it is (once performed) a constant, or fixed, sign. In fact, it is as a fixed time-sign that circumcision signals the relationship of Hebrew males with their God. (Genesis 17:13—"and My covenant shall be in your flesh as an everlasting covenant.") Thus, the Hebrews viewed menstruation as a special sign but rejected it as a Profane, Unclean one while inventing the circumcision-sign as Holy and basically in opposition to menstruation.

Since cyclic time is ever identified with mystic, ecstatic religious consciousness, was it thought that women were thus shamanistic (i.e., possessing special powers) when menstruating? Shamanistic religious consciousness is frequently associated with sickness, which is one of the names for menstruation. Perhaps menstruation was believed to reveal a divine power about women who once may have dictated the life of the human community. Elizabeth Gould Davis theorizes in *The First Sex* that menstruation taboos originated in a previous, great, matriarchal civilization and were designed to protect women and enhance their authority.[21] Neumann speculates that ancient women caring for each other during menstruation and childbirth might "have led to the first blood stanching, healing of wounds, and soothing of pain."[22]

I think it is helpful and important to incorporate such possibilities in comprehending the past and opening a present-future in which we will no longer repudiate and devalue this life experience of menstruation. I sense that menstruation is a powerful occurrence, indeed, a special or "sacred" sign, involved as it is with women's well being and with our self-images. In this connection, it is interesting to hear the words of feminist Robin Morgan, "we are rising, powerful in our unclean bodies ... undaunted by blood we who hemorrhage every 28 days ... we are rising with a fury older and potentially greater than any force in history and this time we will be free."[23]

Such resistance to female oppression has been rising for millenia, even since the time of the ancient Hebrews. The *Torah* tells only the story of male attitudes; we do not have a written record of what it meant to ancient Hebrew women. However, clues to our history do exist in the stories women have always told each other. A young Jewish woman, hearing about my research, passed on to me a fragment of this oral tradition. She said it is an old "rumor" among Jewish women that their ancient foremothers secretly encouraged each other to turn the time of enforced menstrual separation into their own time for rest and meditation; perhaps echoing even more ancient times in which women gathered together for menstrual celebration. The record of menstrual negation is large and has had many consequences, but the actual, living story of women is larger.

## Blood-Sisters

"My flow is coming on, Yana," exclaimed Mave, rubbing her palm in slow circles over her abdomen. "Are you coursing now too?"

"Spoken just a flash before my own announcement, Mave." Yana paused a moment. Crossing the room to where Mave stood, she came up behind Mave, placing her arms around Mave's waist and laying her left palm on top of Mave's slowly moving hand following the circles she made. Yana whispered in her ear "Let's invite Helen over to celebrate with us."

"Certainly! I'm already on my way to the radio-phone." Mave started to pull away slightly to reach out and punch the buttons on their Message Center System. She was dressed in a bright comfortable jumpsuit made of the same soft, scarlet velour as the huge floor cushions scattered around the oval-shaped room. She plopped herself down on one after putting through the call.

"Hello, Helen? This is Mave. I'm calling to invite you over because Yana and I . . . . . . . You *know?* Well, naturally you know. It *is* the new moon and the 13th after all, and Yana and I haven't missed the start of our bloodflow on somewhere within the 12th, 13th or 14th in our last twenty-three cycle revolutions. So, you'll come over and bring some appropriate poems? . . . . . . . Good. See you in about an hour."

                    ❋    ❋    ❋    ❋    ❋

It wasn't really quite an hour before the door bell rang at Mave and Yana's dome-dwelling and Yana ran to open it. "Here I am," said Helen greeting Yana with a hug. "And here is my new friend Seeja, whom I want you and Mave to meet. She's just today arrived here from African Area Three to work for a while in the Planning Council. And . . . we've discovered that *we're* Blood-Sisters." Helen paused, running her fingers through her steel-silver hair, allowing a moment for Yana's pleased murmurs and smiles as she greeted Seeja. "So, of course," she resumed, "that's why I knew you'd be happy for her to come over with me tonight." Yana led Helen and Seeja into the large, round central room where everyone settled down and Yana repeated Helen's introduction of Seeja to Mave. Helen chose an easy chair rather than a cushion and pulled it a little closer to the brick and stone fireplace. She lit a yellow candle selected from a large box of candles on the hearth, reflecting as she did so that she certainly did feel hopeful and happy, in a clear yellow-joy mindmood. She watched as the other three women exchanged greetings.

"Pick your candle," said Mave to Seeja as she pointed to the box which had all colors and sizes of new and old candles.

"Ahmm," smiled Seeja, "I want this dark rich brown one, a little company for me tonight among so many white sisters here." She winked at Mave while taking her candle and lighting it from the fire. She warmed her hands a moment and then settled crosslegged on one of the floor cushions. "You two certainly are dressed appropriately," she remarked, looking first at Mave's scarlet jumpsuit and then at Yana's white and red tie-dyed one. "I too was always inordinately fond of wearing something red when my flow time came 'round."

Mave and Yana looked at each other, a little self-conscious but pleased with Seeja's comments. They were both excited that a new Blood-Sister was here for the flowing and learning Celebration Circle. They quickly began to ask dozens of questions of Seeja, the newcomer.

"Wait, wait, too many questions at once!" Seeja held up her hands in mock protest. "First things first. You want to know how Helen and I found out that we're Blood-Sisters. Well, it's simple really. I was at her house when you called and she told me that she was going to come over here and bring some poems for you all to read aloud and imagine upon. She said you especially liked to invite her since her flowing days had always come between the 12th and the 15th of each lunar month—almost the same interval as you two. So I spoke right up and said that had been my own rhythm, too. And then . . ." Seeja paused a moment, allowing the two younger women to experience a delicious moment of curious anticipation as to what she would say next. "And then, to make things even more strangely special, Helen and I began comparing notes and found that we've each ceased bleeding-flowing and entered star/sun Age/Stage in the same lunar month, eleven years ago. I know there are different patterns of when, and how much, Blood-Sisters harmonize, but *that* really stands out as a coincidence!"

"And as you might guess," added Helen, leaning forwards, "we've spent the day together talking nonstop, and have found quite a few correspondences between us. No wonder we met each other today browsing in the music shop each looking for a recording of the same symphony!"

"Good goddesses!" broke in Yana. "You sound almost identical."

"Oh no, not at all," said Helen. "Besides some more obvious differences," she grinned, reaching her white hand out to hold Seeja's brown one, "we've quite a few noncorrespondences—case in point, we had definite opinions about which conductor's recording of the symphony we preferred, definitely *different* conductors. But . . . " she stopped suddenly, interrupting herself. "Let's hear a poem. I know, Mave, you said you recently found a new one, well actually an old one, from a foresister—back in the twentieth century—and I've been looking forward to hearing it."

"Right," said Mave as she rolled off her pillow and stretched out, her long form making a scarlet stripe across the thick tan rug. She reached across Yana, and picked up a small book, bound in the old-fashioned manner of the 1970s. "This poem is from the time when this knowledge was only a rumor among women that when we are close to each other in psyche (whether or not we are together) we often find that our cycles harmonize and that our blood-flow comes together. In those days, women knew it happened sometimes, but more often than not, it did not happen, or when it did, it was not noticed. Those were the times when the blood was hidden and self-knowledge was stopped up, thwarted, kept separate in the brain-hut and called 'intuition' when it was even called anything at all. This poem is by a woman who dared to start saying what she really knew. It's called 'Falling Off the Roof.' "

"Hold on a minute," asked Yana. She sat cross-legged, her left index finger tracing on her knee the red tie-dye swirls in her jumpsuit. "Let's muse a bit about what such a curious title might mean. I know that the ancient books say it was a flow-name, but such an odd one! Does it come from the time when women were actually put off in separate places while they had their courses flowing? Or is it from the somewhat later times when the blood-flow was called 'the curse' and was said to make us dangerous emotionally and out of control?" She could not suppress a chortle at this last idea. "Out of control!" she sputtered. Yana then began to mock this idea (and to show off her history knowledge to Seeja). "*That* was clearly one of the many lies spread by prejudiced patriarchs! They feared feeling different forms of wisdom so much that they sought to dismiss consciousness-changing rhythms by calling it 'out of control.' "

Helen spoke up quickly. "Yes, they often reversed the truth back then, to try to hide it from women. I think that the name 'falling off the roof' was from the twentieth century of their lord . . . a hard time . . . but a time of upheaval and of women always/evermore seeing. It was a sort of secret shame-name, from times when women did not often say 'flow' or 'blood.' Many people thought rhythm-woven knowing/flowing was a bad unstable danger-weakness!"

At these words, Mave, who had been knotting her brows pensively, sat up straighter and said suddenly, "I've got a flash about it!"

"What, Mave, what? Tell! Tell!"

"Perhaps it had an undermeaning. Moving down or falling from the roof is a dramatic change. Perhaps that's part of it, that image, I mean, of changing. Perhaps it's an unconscious image of changing consciousness." The other three women looked dubious but interested. Mave continued with her thinking out loud undeterred pursuing her new intuition/idea with enthusiasm. "I think it could be a preknowing of the nature of rhythm perception. But masked with a descriptive name that sounds like a *bad* change, an injury, hurt, wounding, fear."

Seeja began to nod slowly in agreement. "Yes, it could be so," she said. "In those days when sacred holy people showed special changes in their consciousness rhythms, they were called sick and thought to be ill or struck down by magic powers. People were afraid of such flowing/knowing."

"Well, maybe," Yana's finger stopped tracing the patterns on her jumpsuit and she looked apprehensive. "It's a name that frightens me a little. Just conjure what it must have been like—women had only haphazard, almost denied, furtive times for any relaxing into the coursing-flowing/knowing. We *are* fortunate to live in an age that respects this as a time of calming, a stage of sensing our mind-body vibrating with a full richness . . . perceiving the subtleties of all things . . . the nuances in word meanings, slight shifts among shades of color, tiny touches of difference in tones of music . . . . " Yana's voice trailed off as she murmured these last phrases.

"The patriarchally-possessed ones called the revelling in knowing nuances 'slow reaction time' and 'off normal performance.' How upside down! They cursed women as slow and clumsy and out of synchronization with their death-machine standard of constant ticking and tocking living. Obviously the real problem was that *they* didn't know how to go smoothly into synchronization with other dimensions of perception." Mave sighed rather heavily. "But, it's easy for me to see that their problem was being monotonously monodimensional since I did *not* grow up learning lies." Mave stopped speaking, her anger at the past oppressive ignorance had been really fully expressed, and now she felt herself moving on and mellowing into a mood of sadness for the pain of the past, and into the thoughtful appreciation of her present situation. She stared into her candle.

"Well . . ." announced Seeja, with a slightly teasing tone, "my appreciation of nuances and rhythms has revolved completely around again to wanting to hear the poem."

"Right," said Mave, finally opening the slender volume of poetry. Everyone wiggled a bit in her seat, settling into a comfortable meditation/attention posture. "Karen Lindsey wrote this," said Mave, as she started to read in a clear, firm voice—

'Falling Off the Roof (for E.C.)

*swollen cotton*
*drops, blood*
*swirls in water*
*red and wet as our mother's first and glorious sin*
*red*
*as the glowing skin of the*
*apple you may not touch*
*wet as the rich flesh*
*of the fruit you may not taste*
*red & wet & i have finally*
*plunged my hand into my bloody cunt*
*licked my fingertips and*
*smeared my face with blood with*
*war*
*paint. . .*

*when i learn to love my blood,*
*the revolution's begun'*[24]

They were all silent for a while. "A brave poem," Helen commented softly. And they sat silently for a while more, listening to the fire, feeling their clothes settled soft against their skin, thinking each to herself of ideas and inspirations that she would sort out and mold into projects and plans when a more linear energy spun around. Seeja was the first to break the silence. "I want to offer a story-gift and tell Mave and Yana something about what it's like to be eleven years into my star/sun age/stage."

"M-m-m-m-m-mm." hummed the two young women in unison. "What do you want to tell?"

"Well *I* don't know . . . that depends on what *you* want to know about it?" Seeja cackled as she said this, pleased with herself at having woven in a little joke on them so smoothly. And by the time she drew another breath, the other three women were cackling too.

"Do you miss rhythm harmonizing?" asked Yana, as she threw some pine cones on the fire.

"I still know rhythms," said Seeja. "You must realize that all living is filled with many moving, cycling rhythms. The bleeding-flowing age/stage is only one sort of phase-changing among myriad changing phases in the self. I do move with a new, more specific constancy throughout my rhythm-changes now . . . but this is certainly not a loss of rhythm knowing. I have in me the memory/experience of about four hundred revolutions of bleeding/coursing. I have had my ears and eyes and nose and brain and tongue and skin—all dimensions of me— tuned well. Those lessons I have learned, and now many rhythms are in me subtly, able to vibrate with recognition. With such a memory depth, I can use the now available constancy of my star/sun age/stage for focus, for finding threads of thinking, playing, living which seem best to add to my self and to the needs of now."

Helen looked steadily at Seeja as she spoke. She knew Seeja's story-wisdom was not completely new information to the two eager questioners. Helen herself had shared with them her way of telling this description about the time past the menstruating age/stage . . . all four knew it . . . , but it was a satisfying thing to chant and rechant respect for each of the ages/stages.

Yana and Mave looked at each other. "We were wondering . . ." began Yana, " . . . we talked about it just yesterday . . . we were wondering if we might pass into the star/sun age/stage at the same time? After all you two did. We didn't enter bleeding-flowing together. Mave was first by almost two lunar years, even though we're the same birth year. But very soon after I did start, we began to harmonize into being blood-sisters."

Before Helen or Seeja could comment, Mave started up as though anticipating their response. "I know. I know. There is change and there is flux . . . and we may well pass on in our transforming to have very different cycles. *But* . . . , we have been in this wave pattern of flow for fifteen years . . . with some lengthening and some shortening here and there. So anyway, perhaps the coursing will stay together for us."

"Ahhhhh, well," began Seeja slowly, stretching her feet out toward the fire and propping them up on a little cushion, "it is clear, as Helen told me, that you two share much and do love each other. But don't romanticize it and try to *grab* on to the harmonizing to make it stay. That would be false, a sharp stopping of living/flowing. Whatever your cycling moves into—let it be. Listen," she said earnestly, her black/brown eyes burning intently, "some things we do not know, for if we did . . . there would not really be change, but only static certainty—absoluteness, killing fixity . . . ." Seeja broke off speaking and looked calmly and intently at Helen, silently summoning forth her friend's words to round out the lesson/lore.

"Remember . . ." began Helen, picking up a pine cone from the hearth basket and tossing it gently but rather suddenly to Yana, "remember *before* you started your bleeding age/stage? You had a delicious, stretching, reaching body-growing. Now your body-growing has sprouted you up and filled you out and you carry the memories of that sort of living. You carried that growing-knowing into your bleeding-flowing rhythms. Back then you were emerging and you didn't know *when* the bleeding cycle years would begin. You knew only that they would. And remember," Helen paused a moment . . . "remember especially how—when the change into a new age/stage occurred—you learned the leaving-lesson? You experienced very clearly the loosing/leaving/ending aspect of transforming." Helen added these last words in a serious tone that was not, however, sad but, rather, soothing.

"Yes," mused Yana, turning over the pine cone in her hands. "I was so happy to begin bleeding and to be discovering myself moving into the second age/stage . . . yet it puzzled me at first that even so I yearned for something that I knew was gone . . . and would not be again. I found in myself then the realization that the mourning for one phase must be in the motion that moves on to the greeting of another. Truly, it is leaving/ending/dying which moves the edge of the circling cycles back again into meeting/beginning/living."

"That contemplation reminds me of a poem I brought tonight," said Helen. "It's by June Namias." She looked through a folder of papers and drew out a red sheet lettered in black ink. She waited while everyone's breathing regulated into each woman's individual even-concentration circulation. Then she began,

'Cycles

*Night*
*last night*
*blood between my thighs*
*my body watches as from nowhere*
*a moon moves across the sky.*

*Daylight*
*beige-tipped leaves — —*
*trees wait patiently for the inevitable.*

*Noon*
*two deer stand*
*beautiful, unknowing*
*on fresh mowed fields.*
*In a month their bodies match the hunter's jacket.*

*Dusk*
*the berry is at midcycle*
*hanging in red thickets*
*at ease.*
*With luck seeds can survive a frozen future.*

*Regeneration is not merely*
*the other side of melted snow.*
*How much death it takes for things to grow.'*

"Oh Helen . . . I didn't know that one," Seeja whispered. "Let me copy it from you before I leave to return to my settlement." They all spoke in whispers for a while. Helen brought out the pomegranate wine she had made several summers ago. Yana added some goat cheese her birth clan had sent her for her birthday. They watched the cats who had been dozing all along in front of the fire and Mave began to play with the kittens, twirling a string above them. Gradually the spell shifted and smiles, then chuckles, then shouts of laughter broke out in response to the kittens' acrobatics.

"Oh! Oh! I've got another poem!" exclaimed Yana with a loud laugh. "I know just the one!"

"You look pleased with yourself," observed Helen affectionately. "What is this poem?"

"Just the one, just the one," Yana repeated jumping up to stand in front of the fireplace, adopting a pretend pompous air. "It continues this evening's cycle . . . since it's also from a foresister and was written in the sisterhood for the same woman as the first poem. See how our women energy courses in and out and right back around! . . . This one has a funny mood. I learned it as a schoolgirl—when we were first taught about the ancient struggle days when patriarchal oppression/ possession made people's lives so stupidly miserable. It was so somber —learning about the ways women and men were warped by conditioning into dualistic opposition. So, the teacher gave us this little poem—which by careful coincidence is also about struggling with school—to cheer us up. I like its happy rebel obstinancy . . . ready? It's by Leigh Star:

'Fascination (for E.C.)

*go away Harvard*
*can't you see I'm busy*
*menstruating?'*

"Ha!" burst out Mave, just as Yana finished the last word—which in fact, she had drawn out into a rather long, droll drawl. "I remember learning that one, too."

"M-m-m," Seeja hummed, "that reminds me of my school times. . . . I remember a special poem from then, too—but, it's not just about menstruation-flowing, but all our blood . . ." her voice trailed off as she considered whether to continue.

"Now I'm intrigued," said Helen. "One more poem seems right to round things out."

"I'm convinced," said Seeja. "This does seem right somehow since I was speaking earlier about the body-memories in star/sun age/stage— and this poem by Yoko Ono is about a deep sweet body-memory." She drew a long breath and all four women looked calmly and directly at each other. Then she said slowly and softly,

*Once we were fish*
*moving freely in the sea.*
*Our bodies were soft and swift*
*and we had no belongings.*

*Now that we crawled out of the sea*
*we are dry and full of cravings.*
*We wander city to city*
*carrying the memory of the sea*
*    (but it isn't just a memory).*

*Listen very carefully and you will hear*
*the sea in your body.*
*You know, our blood is seawater*
*and we are all seacarriers.*[25]

## Notes

1. For more information on Menopause, *see* article by Marlyn Grossman and Pauline Bart in this volume.
2. Such as, Karen Paige, "Women Learn to Sing the Menstrual Blues," *Psychology Today,* September 1973.
3. Estelle Ramey, "Men's Cycles (They Have Them Too, You Know)," *The First Ms Reader* (New York: Warner, 1972).
4. See also my paper on "Zoroastrian Menstruation Taboos: A Women's Studies Perspective," *Women and Religion: 1972 and 1973 Revised Edition* (Missoula, Montana: American Academy of Religion and the Scholar's Press, 1972).
5. A longer version of this paper was originally presented at the American Academy of Religion, New England Area Meeting, Spring 1973.
6. A. Cohen, ed., "Leviticus," *The Sencine Chumach* (London: The Sencine Press, 1967).
7. *Ibid.,* 17:11.
8. *Dictionary of the Bible, Volume IV* (New York: Charles Scribner's Sons, 1902), p. 825.
9. *Ibid.,* p. 826.
10. *Ibid.,* in a footnote on p. 828.
11. The Unclean and the Holy did not function as direct opposites. The not-Holy was simply common, and the not-Unclean was clean. The ordinary world, the arena of human living, was a middle ground of the clean and the common, with Sources of both Holiness and Uncleanness. (Thus, to purify oneself from an Uncleanness did not immediately make one holy but only clean and hence fit to approach the Holy.) Holiness was Divine Power and came from Yahweh (God). Uncleanness was Profane Power and came directly from a Source (called "fountain") which was in itself Unclean.
12. Herbert Danby, trans., *The Mishnah* (Oxford: Clarendon Press, 1933), p. 803.
13. Danby, *Mishnah,* Teheroth, Niddah, p. 747.
14. Since analogy is an important principle of interpretation in the *Torah* and the *Mishnah,* to disallow it in this case is significant.
15. Charles B. Chavel, trans., *The Commandments, Volume 1 of Moses ben Maimon* (London: The Sencine Press, 1967), p. 87.
16. Raphael Patai, *Sex and Family in the Bible and the Middle East* (New York: Doubleday and Co., 1959), p. 158.
17. Philip Blackman, ed., *Mishnayoth, Volume VI, Order Taharoth* (New York: The Judaica Press, 1964), p. 591.
18. Personal interview with Rabbi Gold and printed copy of interview with Rabbi Poupke.
19. Personal interview with Ms. P. Greenfield.

20. Nathaniel Nicklem, *The Interpreter's Bible, Volume II* (New York: Abingdon Press, 1953), p. 53.
21. Elizabeth Gould Davis, *The First Sex* (Baltimore: Penguin Books, 1972), pp. 90–92.
22. Erich Neumann, *The Great Mother,* translated by Ralph Manheim (Princeton: Princeton University Press, 1963), p. 290.
23. Robin Morgan, "Good-bye to All That," *Voices from Women's Liberation,* edited by Leslie B. Tanner (New York: Signet Books, 1970), pp. 275–76.
24. Karen Lindsey, *Falling off the Roof* (Cambridge, Mass.: Alice James Press, 1975), p. 41.
25. Yoko Ono, *Grapefruit* (New York: Simon and Schuster, 1971).

## Annotated Sources

Weideger, Paula. *Menstruation and Menopause: The Physiology and Psychology, the Myth and the Reality.* Knopf, 1976.
As the comprehensive title suggests, this book is a thorough treatment of menstruation and menopause, discussing health information, historical attitudes to menstruation, and much more. It is a well written readable book and, to my mind, is the best one available. It is written from an open-minded, feminist perspective.
Delany, Janice; Lupton, Mary Jane; and Toth, Emily. *The Curse: A Cultural History of Menstruation.* Dutton 1976.
This book concentrates more thoroughly on attitudes, customs, art, and humor about menstruation. It is a fascinating analysis of much diverse and interesting information.
Boston Women's Health Book Collective. *Our Bodies, Ourselves: A Book By and For Women* (Revised & Expanded), Simon and Schuster, 1976.
This book has become a classic feminist introduction to women's health information. Chapter 2 has a clear explanation of how the menstrual cycle functions.
_____. "Menstruation."
A recent 8-page pamphlet with general information about menstruation, keeping your own records, using a sponge, etc. This pamphlet contains a collection of remedies for menstrual problems. For a free individual copy, send a self-addressed, stamped envelope to: Boston Women's Health Book Collective, Dept. B., Box 192, West Somerville, MA 02144.
Culpepper, Emily. *Period Piece.*
A 10 minute, 16mm color film exploring attitudes, stories, and images of menstruation, concluding with a woman's first vaginal self-examination during her period. For more information write to: Culpepper,

64R Sacramento St., Cambridge, MA 02138; or Insight Exchange, P.O. Box 42594, San Francisco, CA 94101.

Wheat, Valerie. "The Red Rains: A Period Piece." In *Chrysalis: A Magazine of Women's Culture,* No. 1., Spring 1977.
This article contains a collection of resources for information on menstruation and body rhythms. In addition to the sources listed here, it contains some more general references.

Montaingrove, Jean and Mountaingrove Ruth, eds., *Woman Spirit Magazine,* Vol. 1, Winter Solstice 1974. Box 263, Wolf Creek, Oregon 97497.
This issue in particular has several articles and poems about menstruation, including new feminist rituals and customs. Other issues of *Woman Spirit* often contain personal essays and poems about menstruation, especially considerations of menstruation as sacred power.

[Emily E. Culpepper *is a rebellious white woman, originally from Georgia, now living in Cambridge, Massachusetts, where she is a doctoral student at Harvard. She is immersed in work on her dissertation on* Feminist Myth-Making: A Philosophy and Force for Social Change, *and has taught in a course on Folklore and Mythology at Harvard University. She has published numerous articles in academic and feminist journals. Her most recent publication is a chapter entitled "The Spiritual Dimensions of Radical Feminist Consciousness" (in* Understanding the New Religions, *edited by J. Needleman [New York: Seabury Press, 1978]). Her longstanding fascination with menstruation has led her to collect a variety of women's stories, myths, anecdotes, artwork, health information, and customs. She is currently writing with Harriet Arnoldi a* Menstrual Playbook for Girls of All Ages. *She is a founder of Artemis Productions, a lesbian group that produces women's music. She hopes feminists will explore class and race a lot more (having always worked to support herself), feels blessed to have supportive women friends, and is passionately committed to the cause of women.*]

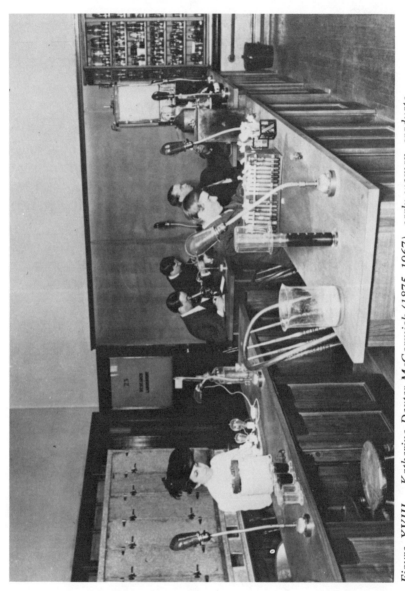

*Figure XVIII. Katherine Dexter McCormick (1875–1967), early woman graduate of M.I.T. (B.S. in biology, 1904). (M.I.T. Historical Collections)*

# Marlyn Grossman and Pauline B. Bart

---

# Taking the Men Out of Menopause

*They call me Grace.*
*Yesterday I went*
*to the grocery store.*
*I had filled up*
*the cart*
*and was half way through*
*the check stand*
*before I realized*
*I had shopped for the whole family.*
*The last child left*
*two years ago.*
*I don't know what*
*got into*
*me.*
*I was too embarrassed*
*to take things back*
*so I spent the week cooking*
*casseroles.*
*I feel like one of those*
*eternal motion machines*
*designed for an*
*obsolete task*
*that just keeps on*
*running.*

> —Susan Griffin, *Voices*

Menopause, the cessation of menstruation, is the second "change of life" that we go through as women.[1] In our society, with its emphasis on youth, this is an unappreciated, often maligned time; there is no bar mitzvah for menopause. As Ursula LeGuin said, "It seems a pity to have a built-in rite of passage and to dodge it, evade it, and pretend nothing has changed. That is to dodge and evade one's womanhood, to pretend one's like a man. Men, once initiated, never get the second chance. They never change again. That's their loss, not ours. Why borrow poverty?"[2] Despite the intrinsic appeal of LeGuin's position, there have been many forces at work pressing women to see themselves differently. It is no accident that the powerful male-dominated institutions of our society, particularly medicine, have functioned here as in so many other cases to define the ways in which women are different from men as deviant, diseased or, at the very least, undesirable.

The topic of menopause seems particularly to evoke such sentiments. This may derive from the conviction with which a nineteenth-century medical "authority" like the prominent gynecologist Charles D. Meigs could describe woman as "a moral, a sexual, a germiferous, gestative and parturient creature."[3] Another physician, Holbrook, pontificated that it was "as if the Almighty, in creating the female sex, had taken the uterus and built up a woman around it."[4] Historian Peter Stearns observed that in eighteenth- and nineteenth-century Europe physicians thought women decayed at menopause.[5] Victorian physicians invariably characterized it as the "Rubicon" in a woman's life, and medical popularizers of the day "blamed the frequency and seriousness of disease during this period upon the 'indiscretions' of earlier life."[6] Kellogg remarked that the woman who transgressed nature's laws will find menopause "a veritable Pandora's box of ills, and may well look forward to it with apprehension and foreboding."[7]

Since then, physicians' attitudes have not changed much. However, it might be surprising to some not familiar with the pervasive sexism in current gynecological writing,[8] or with the traditional anti-female ambiance in medical education,[9] to learn that at a conference on menopause and aging sponsored by the U.S. Department of Health, Education and Welfare—a conference uncontaminated by the presence of a single female participant—Johns Hopkins' obstetrician-gynecologist, Howard Jones, characterized menopausal women as "a caricature of their younger selves at their emotional worst!"[10,11] To quote Mary Brown Parlee, "It sometimes seems as if the only thing worse than being subjected to the raging hormonal influences of the menstrual cycle is to have these influences subside."[12]

Physicians are willing to prescribe "for the menopausal symptoms that bother him [the husband] most,"[13] even though the drugs may be of questionable value and can have harmful side-effects. Estrogen is the one most frequently prescribed. It has been touted as effective in controlling menopausal symptoms from general ones such as depression to specific ones such as hot flashes (which are nervous system responses triggered by lower estrogen levels). A gynecologist named Wilson (whose research, not coincidentally, was sponsored by a drug company) warned women that to stay "Feminine Forever," they should take estrogen as long as they live, because menopause is a "deficiency disease."[14,15] This redefinition of a natural event such as menopause into a disease is an example of the increasing medicalization of normal events in our lives. Childbirth is another example of an event—once considered a natural part of women's lives—that the medical establishment now treats as pathological.[16] This process is both a cause and an effect of the enormous power American physicians have to define and manipulate our reality.

Actually, Wilson's skill as salesman far exceeded his accomplishments as a medical researcher.[17] Over 300 articles promoting estrogen have appeared in popular magazines in the intervening years. Yet researchers started reporting increases in uterine and breast cancer in women taking estrogen as far back as the 1940s,[18] and almost twenty years ago, animal studies started showing that estrogen can induce cancer in estrogen-dependent organs (breast, uterus, cervix, and vagina).[19] More recently, well-designed studies have shown that the risk of developing cancer of the lining of the uterus is four and one-half to fourteen times greater for women who take estrogen than for women who don't, and that the longer a woman takes estrogen, the greater her risk becomes.[20,21] Another study showed a possible link between taking estrogen and developing breast cancer.[22]

Ayerst, the drug company that has been grossing about seventy million dollars a year from Premarin, the estrogen most frequently prescribed to menopausal women, took all this in its stride. Immediately after publication of the 1975 studies, Ayerst sent physicians a letter (which did not even mention the cancer studies) recommending "business as usual." Alexander Schmidt, then Food and Drug Administration Commissioner, called this act "irresponsible."[23] Later, Ayerst hired the public relations firm of Hill and Knowlton, which specializes in companies with image problems, to help them deal with their increasingly bad press. In a letter dated December 17, 1976, a Hill and Knowlton vice president recommended to Ayerst an impressively complete and cynical media campaign "to protect and enhance the identity of estrogen replacement therapy."[24]

Even such high-powered planning may not be able to save Ayerst's business entirely, though. On July 22, 1977, the Food and Drug Administration proposed a regulation that estrogen drugs shall contain patient package inserts detailing what estrogen does and does not do. The model insert that the FDA prepared for comment is unusually candid and direct: "You may have heard that taking estrogen for long periods (years) after the menopause will keep your skin soft and supple and keep you feeling young. There is no evidence that this is so, however, and such long-term treatment carries important risks." Further, "Sometimes women experience nervous symptoms or depression during menopause. There is no evidence that estrogens are effective for such symptoms and they should not be used to treat them. . . ."[25] The model insert also contains several other hard-hitting statements.

Before they become effective, federal regulations must be published to allow time for public review and comment. After a set period has elapsed, a final version of the regulation is published and becomes law in sixty or ninety days unless someone petitions the relevant agency and/or federal court to prevent its acceptance. Not surprisingly the drug companies, represented by the Pharmaceutical Manufacturer's Association, have both petitioned the FDA and sued in federal court to block the estrogen regulation. What is less expected and more distressing is that the American College of Obstetricians and Gynecologists joined in the suit. Subsequently, the American Pharmaceutical Association, the American Medical Association and the National Association of Chain Drug Stores have also lent their support to the court action.

There's something for everyone in menopause: patients for physicians, profits for drug companies, and cancer for women. The medical and pharmaceutical groups are claiming that including the information brochure in the package will interfere with the traditional doctor-patient-pharmacist relationship (which does seem possible—though perhaps in ways that will be of benefit to the patient!). In a predictably paternalistic tone, the American Pharmaceutical Association claims, "The officially composed leaflet is far from understandable—to many patients, it will be utterly incomprehensible. And much of the information mandated for the leaflet is not only in no way pertinent to *the proper concerns of the patient* (emphasis added), once therapy has been determined, but it is wholly unsuitable for lay persons without medical or scientific training."[26] (If you feel as if your intelligence has just been insulted, you are in good company.) Finally, the medical and pharmaceutical groups claim that the FDA has acted without legislative authority.

The following excerpt of an article entitled "DOCTORS' GROUPS FILE THREE SUITS: FDA's estrogen ruling fought" that appeared in *The Boston Globe* on 9 October 1977, perhaps best reveals the nature and extent of the medical establishment's opposition to the FDA action:

"Indeed," said Dr. H.J. Barnum Jr., owner of a public relations firm in New York, in a written affidavit, "it is hard to imagine a class of patients more susceptible to adverse psychological reactions than the menopausal female, the very target of this leaflet, and nearly every aspect of the [pamphlet] appears calculated to arouse an emotional reaction in the patient."

"The best the [pamphlet] will accomplish," added Detroit Gynecologist T. N. Evans in his affidavit, "is a massive scare which the medical evidence indicates is wholly unwarranted." Evans added that the patient could even experience such an erosion of confidence "that she would not bother contacting her doctor at all, but simply discontinue consultation with the physician in the mistaken[!] belief that he lacks competence or adequate concern for her."

On the other hand, the Detroit doctor said, so many women might call their doctors after reading the pamphlet that this would put "an additional strain on an already overtaxed health care system."

Another issue in the suits, somewhat more in the background, is the fear of both drug companies and doctors that women who read the pamphlets may be more inclined to sue for malpractice, claiming that uterine cancer or some other serious disorder was caused by estrogen pills or their misuse.

As this is being written (October 18, 1977), the FDA regulation has become law and all packages of estrogen are now required to contain a patient information brochure. Though the medical and pharmaceutical groups have lost their attempt to get a temporary injunction against the enforcement of the regulation, they continue to press their suit seeking a permanent injunction. Since the judge who ruled in the temporary injunction case cited the substantial evidence of risk of cancer and overuse, it seems unlikely that the permanent injunction will be won. (Indeed, the claim that the FDA is acting beyond its legislative authority is belied by the fact that there have been patient package inserts in oral contraceptives since 1970).[27] Nevertheless, these actions of the medical and pharmaceutical groups dramatize the sexism and general inhumanity of the male dominated, profit-oriented U.S. medical system. A "deficiency disease" was invented to serve a drug that could "cure" it, despite the suspicion that the drug caused cancer in women.[28] That the suspicion had been voiced for so many years before anyone chose to investigate it is yet another example of how unimportant the well-being of women is to the men who control research and the drug companies who fund much of it. And the unwillingness of physicians, pharmacists, and drug companies to give women the information now available about estrogens demonstrates once again that the powerful will not give up any of their power (to say nothing of their financial gain), even after it has been clearly shown that they are using that power to harm women.[29]

We also have the male medical establishment to thank for the paucity of information about menopause. (It is difficult to imagine such ignorance about an event in the life of every man.) Sonja and John McKinlay surveyed what little literature there is about menopause and found it wanting.[30] Much of it is based only on physicians' "clinical experience," which is notoriously selective and unreliable. Where more objective research has been attempted, it has frequently involved retrospective data (which introduces all the unreliability of memory), unclarified cultural differences in recognizing and reporting symptoms, and the use of non-uniform definitions of menopause and of its symptoms.

What little we know about how *most* women (that is, those who do not end up in doctors' offices) experience menopause comes, not surprisingly, from research done by women.[31] Sonja McKinlay and Margot Jeffreys surveyed over six hundred women between the ages of 45 and 54 in and around London.[32] Most women go through menopause around 50. Three-quarters of the menopausal women were experiencing hot flashes. One-quarter of the post-menopausal women continued to have hot flashes for five years or more. The six other symptoms that were specifically inquired about—headaches, dizzy spells, palpitations, sleeplessness, depression, and weight increase—were each reported by one-third to one-half the women. While having these symptoms was not related to whether a woman was pre-, post-, or currently menopausal,[33] a woman who reported any symptom was likely to report more than one. Neither employment outside of the home nor the work load within it were related to the experience of those problems. Despite the fact that three-quarters of the women who had hot flashes found them embarrassing or uncomfortable, only about one-fifth sought medical treatment.

Since professionals have offered women so little information about the menopause, women's self-help groups have done some of their own research. Two groups, one in Seattle and one in Boston, used mail-in questionnaires in an attempt to survey women's physical and emotional experiences of the menopause.

The Seattle group, calling itself "Women in Midstream," had originally set out to investigate what the experience of menopause was like for women who were middle-aged before estrogen replacement therapy was available. Accordingly, they sent one thousand questionnaires to nursing homes, but received only seventy replies. These older women were also unwilling to talk about the subject in face-to-face interviews. This experience is of interest because it shows the extent to which women are socialized to regard normal bodily processes and life experiences with shame and to hide them from public scrutiny. It also suggests that most, if not all, earlier studies of menopause may well suffer from the respondents' unwillingness to reveal to researchers the full extent of their actual feelings and experiences. It may well be that only now, when support is available in our culture for women to share these formerly private areas of their lives with other women, can we really learn about these experiences in a systematic way.

Clearly, some women are now eager to share knowledge about menopause. The "Women in Midstream" group has received more than seven hundred completed questionnaires from highly motivated women who wrote or called to request the questionnaire after they had learned of it in the newspaper or on the radio. Unfortunately, sufficient woman-power and resources have so far been available to analyze only 250.

Because of the method by which they obtained their sample, the "Women in Midstream" researchers think that the respondents are probably largely middle and upper class and have had a relatively difficult experience with their menopause. (Even so, half the group described it as "easy" or "moderately easy!") About 60 per cent come from the state of Washington and the rest from elsewhere in the country. (Most of the sample came from a newspaper column directed to middle-aged women that originates in the *Des Moines Register* but is nationally syndicated in smaller towns and cities.) The vast majority of the women in the sample are white, married, have had about three children, but live with only their husband at this time. The women range in age from 28 to 73, with two-thirds of the group between 45 and 55. 60 percent are Protestant and 18 percent Catholic. Half of the women consider themselves "in the midst" of menopause and one-sixth "all through." Two-thirds of the women are working outside the home.

One of the most striking findings in this group is that three-quarters of the women had been prescribed hormone (i.e., estrogen) therapy. Yet no more than 60 per cent of the group had sought physicians' help for hot flashes and/or thinning and drying of the vaginal walls, the only two menopausal conditions for which there is some agreement that estrogen treatment is effective. (Because doctors have not changed their prescribing habits since publication of the studies that clearly showed the link between estrogen therapy and increased risk of cancer, we would not expect different results were the survey re-run today.) A staggering 55 per cent of the group were prescribed tranquilizers, causing one to wonder how much of this was in response to the women's needs and how much in response to those of their husbands and/or physicians,[34] especially since psychiatric therapy was recommended for less than 10 per cent of the women. Fifteen per cent of the women received dietary supplements. (Obviously, some women received more than one recommendation.) Only 11 per cent were told that they needed no treatment.

Only slightly more than half of the women reported satisfaction with their doctor's attitudes and found her or him helpful. Of those among the group who sought help from non-medical sources, three-fifths found these people helpful. More than 75 per cent of the women discussed their menopausal problems with female friends or relatives and nearly half of them found this helpful. More than 66 per cent of the group discussed their problems with husbands, male friends or relatives and, again, nearly half found it helpful to do so.

The "Women in Midstream" group feels that social supports are very important for women going through menopause. Accordingly, they asked their sample if they would be interested in talking with other women about the health and social problems of older women. More than half the women were definitely interested and another sixth said they might be. The same total number of women indicated definite or possible interest in individual discussions, though more were uncertain here. (The desirability of such discussions is certainly borne out by an incident Paula Weideger reports, in which a woman participating in a menopause consciousness-raising group was surprised to find that while she had originally experienced her hot flashes as uncomfortable, they had now become pleasurable!)[35]

The Boston Women's Health Book Collective attempted an even more ambitious survey than the Seattle women. They set out to study the attitudes toward and experiences of menopause by women of all ages and in all parts of the country. Mail questionnaires were sent to friends and relatives of the Collective's members as well as to all the clinics and counseling centers that ordered the 1973 edition of *Our Bodies, Ourselves*. Replies were received from almost five hundred women. Most of them live in large northeastern or middle-Atlantic cities or suburbs. Less than two-thirds are married (compared with four-fifths of the Seattle group), though another fifth are divorced or widowed. Less than half of the group are Protestant, one-quarter Jewish, and one-fifth Catholic. As a group, they have slightly smaller families than the Seattle sample. Just over a third of the total group was menopausal or post-menopausal. As was the case in the Seattle sample, two-thirds of this group was employed outside the home.

Just over three-fifths of the Boston menopausal women received estrogen replacement therapy. While this is less than the Seattle group, it is still surprisingly high for a group that probably is experiencing many fewer menopausal "problems" (given the ways the respective samples were gathered). Five-sixths of the women talked with friends about their menopausal experiences, while over two-thirds talked to husbands (just about as many as in the Seattle group, though the questions asked were not exactly the same). More than two-thirds talked to children or other relatives and one-sixth each to therapists and women's groups. Two-thirds of the women had friends going through menopause at the same time. Two-thirds (largely, though not exactly, the same group) also received emotional support.

Where the physical experience of menopause blends into the psychological, we enter an area of many assumptions and few facts. Earlier, we quoted some choice descriptions medical authorities have given of the menopausal woman. It is not surprising then that psychoanalysts, who are usually physicians, have also tried to use their "science" to keep women narrowly contained within the roles of wife and mother. Since psychoanalysts follow Sigmund Freud, they tend to describe human development in terms of his phallocentric perspective. Helene Deutsch, who unfortunately demonstrates the fact that women in male-dominated professions sometimes internalize that perspective (frequently as a survival technique), described the menopausal woman as having "reached her natural end—her partial death—as servant of the species."[36] In fact two men, one of them a psychoanalyst, have managed to provide a clear view of this muddy area. They have summed it up this way: "It becomes obvious that a type of confused, fuzzy, and prejudicial thinking has existed in the minds of psychologists, psychiatrists, and physicians in general toward the female. Whereas it is clearly recognized that male psychology must be differentiated from male biology, with an awareness of the effects of one upon the other, no such differentiation has been allowed for females. Males are not governed by biologic maturation processes related to aging. They can be abstract, external, and worldly, and are concerned with jobs and events of the day. For whatever reasons, little attention has been given to the obviously similar dichotomy in the psychology and biology of women. Emotions and cognition both have been viewed as being a part of, and controlled by, biology. Almost all research has proceeded from this obviously muddled hypothesis."[37]

Some fuzzy thinking about women derives from the work of the psychoanalyst, Theresa Benedek. In one article, Benedek commented that menopause brings a diminution of the part "of the integrative strength of the personality which is dependent upon the stimulation by gonadal hormones"![38] In another, she found that there was a drop in hormone levels in the pre-menstrual phase of the menstrual cycle, which also seemed to be marked by "narcissistic regressive preoccupations" (which is psychoanalytic-ese for a woman spending more of her energies thinking about herself than our culture thinks she should!). By an impressive deductive leap, Benedek concluded that the declining hormonal capacity of menopause leads to a diminished capacity to love.[39] The methodological problems in Benedek's research are extensive. Among other things, the sample includes only women undergoing psychotherapy.[40]

Another group of researchers who have attempted to understand women's experience of menopause are social psychologists. Bernice Neugarten and her students (also women, demonstrating again that anyone interested in menopause is likely to be female in gender even if not in perspective) used an attitudes checklist to study women's feelings about menopause.[41] They found that women who had experienced menopause were less likely to consider it a significant event than were pre-menopausal women. Also, upper-middle-class women in particular denied the significance of menopause in a woman's life. While it is good that these studies questioned a wider group of women than just those who turned up in a doctor's office, the studies were done in the early 1960s, and thus before the Women's Movement gave women permission to "speak bitterness"[42] about the limitations of their socially prescribed roles.[43]

Some of the interesting findings which come out of these studies result from subdividing the women interviewed. For instance, Lenore Levit found no difference in general in the amount of anxiety menopausal women suffered compared with post-menopausal women.[44] However, women who were very much invested in their role as mothers were more anxious during menopause than they were afterwards. Thus, the transition appears harder for women who are giving up a role they highly value. (This was true whether or not the children were living at home.) Similarly, whereas middle-class women did not differ in anxiety during or after menopause, working-class menopausal women were more anxious than those post-menopause, possibly because working-class life offers women fewer rewards besides children.

Ruth Kraines studied one hundred women, one-third in menopause, one-third pre-menopause, and one-third post-menopause.[45] Using interviews, self-report forms, and checklists containing many symptoms, some of which are traditionally associated with menopause, she did not find appreciable differences among the three groups in their assessment of their own physical state (although menopausal women did check more of the symptoms particularly associated with the menopause). She did find that women who were low in self-esteem and life satisfaction were most likely to have difficulty during menopause, and that the relationship appeared to be circular, *i.e.,* low self-esteem led to difficulty during menopause which in turn led to low self-esteem. Kraines also found a continuity between a woman's previous reactions to bodily experiences (such as health problems, menstruation, and pregnancy) and her reaction to menopause. She concludes that, contrary to the medical studies, menopause in itself is not experienced as a critical event by most women. She suggests that women who seek medical help differ from most middle-aged women in their physical and emotional reactions to stress.

Medical people tell us that a woman undergoes extensive hormonal changes at menopause and therefore requires (their kind of) help; at the same time, social psychological surveys suggest that most women do not regard menopause as a difficult stage of life. The problem with much of the available literature is that physicians have generalized from the relatively small group of women who seek medical or psychiatric help. And social psychologists, working at a time before there was support in the culture for a woman to express her dissatisfaction with the lack of alternatives open to her after her children are grown, may also have come up with a biased picture. How, then, can we figure out what is really happening to women and how many physiological and sociocultural factors influence the experience of menopause?

One way to tease out the sociocultural from the physiological is to look at cross-cultural studies. Both Nancy Dowty and one of us (Pauline Bart) have studied menopause in this manner. Nancy Dowty worked in Israel studying five sub-cultures which she arrayed on a continuum of modernization, from traditional Arab women at one end to European-born Israeli women at the other, with Jews from Turkey, Persia, and North Africa in between. She found no linear relationship between social change and difficulty during the menopause. The transitional women, midway between traditional life styles and modernization, suffered most: they had lost the privileges afforded traditional women, while not receiving those benefits that modernization confers upon women. They had the problems of both groups but the advantages of neither.

Pauline Bart, using the Human Relations Area Files as well as ethnographic monographs, found that certain structural arrangements and cultural values were associated with women's changed status after the childbearing-years.[47] These are summarized in the following table.

*Table I. Characteristics of societies in which women's status changes at menopause.*

| Status Rises | Status Declines |
| --- | --- |
| Strong tie to family of orientation (origin) and relatives. | Marital tie stronger than tie to family of orientation (origin). |
| Extended family system. | Nuclear family system. |
| Reproduction important. | Sex an end in itself. |
| Strong mother-child relationship reciprocal in later life. | Weak maternal bond; adult-oriented culture. |
| Institutionalized grandmother role. | Non-institutionalized grandmother role; grandmother role not important. |
| Institutionalized mother-in-law role. | Non-institutionalized mother-in-law role; mother-in-law doesn't train daughter-in-law. |
| Extensive menstrual taboos. | Minimal menstrual taboos. |
| Age valued over youth. | Youth valued over age. |

It seems reasonable to assume that an increase in status would increase the likelihood of feelings of well-being, so that even if physiological stresses are experienced at menopause, they are well buffered. This appears to be the case in kinship-dominated societies. When this system begins to break down, as it is now doing in some Third World countries, problems arise similar to those faced by some women in our culture. For example, one (Asian) Indian mother who brought up her children to live in the modern manner, that is, independently, felt very lonely and commented: "I sometimes feel, 'What is the use of my living now that I am no longer useful to them [her children]!' "[48]

It is easy to see why some women in our society find middle age stressful. Except for the mother-child bond, which in our society is strong but *non*-reciprocal, we fall on the right hand side of the table, with the cultures in which women's status drops in middle age. It is true that for women whose lives have not been child-centered and whose strong marital tie continues, or for those whose children set up their own residence near the mother, the transition to middle age may be buffered. However, for women who have emphasized the maternal role or the glamor role, middle age may be difficult. Our emphasis on youth (particularly in women) and the stipulation that mothers should not interfere in the lives of their married children (the mother-in-law syndrome) can make middle age stressful for women who have not had the opportunity to invest themselves in anything besides wife and mother roles. By examining the question in a cross-cultural perspective, however, we can observe the multiplicity of *possible* roles for middle aged women and appreciate the fact that middle age need not be a difficult time. Indeed, it *can* and *should* offer women its own unique rewards.

While the cross-cultural evidence strongly suggests that the phenomenon of menopausal depression is related not to physiological changes but to social and cultural structures and factors, a study of individual, depressed, menopausal women in our society could shed a great deal of light on how these factors operate in American culture. One of us[49] has studied the hospital records of over five hundred women between the ages of forty and fifty-nine who had had no previous hospitalization for mental illness. The records were drawn from five mental hospitals ranging from an upper-class private institution to two state hospitals, and were used to compare all the women diagnosed as depressed (whether neurotic, psychotic involutional, or manic depressive) with those who had received other diagnoses. Twenty intensive interviews were also conducted to round out the picture obtained from the records. The interviews included questionnaires used in studies of "normal" middle aged women, and a projective biography test consisting of sixteen drawings of women at different stages of their life cycle and in different roles.

Statistical analysis of the hospital records indicated that depression was associated with current role loss and even with the prospect of loss. Housewives were particularly vulnerable to the effects of losing roles such as those of wife or mother. Ethnicity is another relevant variable, with Jewish women showing the highest rate of depression. When all women having overinvolved or overprotective relationships with their children are compared with women who do not, however, the ethnic differences almost wash out. (Thus you do not have to be Jewish to be a Jewish mother, but it helps a little.) Overall, the highest rate of depression was found among housewives who had overprotective or overinvolved relationships with their children who were currently or soon leaving home. Thus the lack of meaningful roles and the consequent loss of self-esteem, rather than any hormonal changes, seemed largely to account for the incidence of menopausal depression.

This hypothesis received further support from the results of the interviews. All the women with children, when asked what they were most proud of, replied "my children." None mentioned any accomplishments of their own, except being a good mother. When asked to rank seven roles available to middle-aged women in order of importance, the mother role ("helping my children") was most frequently ranked first or second. When children leave home, however, the woman is frustrated in attempting to carry out this role she values so highly and suffers a consequent loss of self-esteem. It is precisely to the extent that a woman "buys" the traditional norms and seeks vicarious achievement and identity (which society has told her is the appropriate route to "true happiness" and "maturity") that she is vulnerable when her children leave. Moreover, many of these women then experience their life situation as unjust and meaningless because the implicit bargain they had struck with fate did not pay off. In the words of two of the women:

> I'm glad that God gave me ... the privilege of being a mother
> ... and I loved them [my children]. In fact, I wrapped my love
> so much around them ... I'm grateful to my husband since if it
> wasn't for him, there wouldn't be the children. They were my
> whole life ... My whole life was that because I had no life
> with my husband, the children should make me happy ... but
> it never worked out.

> I felt that I trusted and they—they took advantage of me. I'm
> very sincere, but I wasn't wise. I loved, and loved strongly and
> trusted, but I wasn't wise. I—I deserved something, but I thought
> if I give to others, they'll give to me. How could they be different?
> ... But, you see, those things hurt me very deeply, and when I had
> to feel that I don't want to be alone, and I'm going to be alone,
> and my children will go their way and get married—of which
> I'm wishing for it—and then I'll still be alone, and I got more and
> more alone, and more and more alone.

Statements such as these poignantly portray middle age as a time when reality overtakes women's dreams for the future and some women are confronted with the meaninglessness of their lives. Women have been taught to believe that they can achieve "true happiness" through self-abnegation and sacrifice for their husbands and their children, that to do anything for themselves is selfish. Some women are able to evade or overcome this script. Some continue to receive the vicarious gratification they sought—their husbands are still alive, well and attentive; their children have made "proper" marriages and/or embarked upon "proper" careers; grandchildren have arrived; and significant others congratulate them on a job well done. But for others, the story is different. They are bewildered by their children's life style, which rejects the values the mothers have worked so hard to attain, indeed not so much for themselves as for those very children. They cannot understand why their daughters do not want to have the children they were taught were their destiny, thus denying them the grandchildren they so joyfully anticipated and who would give new meaning to their lives. Their husbands may have left them for younger women to bolster their own waning egos and diminished potencies (on average, the second wife is younger than the current age of the first wife).[50] For these women to seek a younger mate would be thought ludicrous, since they are no longer considered sexual beings (though of course physiologically, unlike men, they are as capable of sex as they ever were).[51] And because so many of these events coincide with menopause, their effects are attributed to this physiological change.

Alice Rossi, analyzing recent census data, notes that maternity has become a very small part of the average adult woman's life: a woman who marries at twenty-two, has two children two years apart, and dies at seventy-four, on average will spend one-quarter of her adult life without a husband, two-fifths with a husband but no children under 18, one-third with spouse and at least one child under 18; but only one-eighth of her life in full-time maternal care of pre-school age children.[52] This projection dramatizes the inappropriateness of the standard script. It is important that women learn this message, and early, so they will not become casualties of the culture in middle age.

As feminists, we believe that societal problems cannot be dealt with on an individual basis: there are no individual solutions except for those few women who may slip through by chance or special privilege. (It is no accident that Bernice Neugarten's view—that middle age is a neutral and frequently almost positive experience—relies heavily on a study of middle-aged people who are elite professionals or in business and for whom terms like autonomy, predictability, and choices have meaning, particularly at this stage of the life cycle.[53] As the saying goes: "Rich or poor, as long as you have money!") For most women's lives to change, sweeping economic and social reforms are essential. For the present, the only changes we can count on are those that can be brought about by the organized efforts of many women working together to structure alternatives for themselves and for others. While true long-range solutions would require changes in women's lives from very early ages (and many women are actively working at bringing these about), there is growing support for women already in their middle years. The National Organization for Women (NOW) has a task force on older women. There are also increasing numbers of rap groups for middle-aged women both here and in other countries.

For instance, there are menopause rap groups throughout Holland. Since few Dutch married women work outside their homes, the usual problems of transition are exacerbated by the lack of alternative sources of self-esteem. While the women originally join the groups to discuss problems around menopause, broader life style and life cycle issues emerge. In the supportive atmosphere of the groups, many women later are able to express sexual or marital dissatisfactions.[55]

One of the signs of the successful impact of the Women's Health Movement is the fact that gynecological self-examination, once a revolutionary cry of a small group, has begun to be a part of routine office practice among some of the more forward-looking gynecologists. As women demand more participation in, and control over, the various elements of their lives, we expect that the heavy taboos on bodily functions (and what woman has not been anxious lest she "stain through" during her menstrual cycle?) will decrease, so that the embarrassment caused by the hot flashes most menopausal women experience can be alleviated without the use of cancer-causing drugs.

There is much talk of the wider range of options available to middle-aged women. Group support can enable those women who have greater options to use their new freedom to change their life styles and fulfill some of their deferred dreams. However, options are limited by economic conditions, racism and previous educational opportunities. Only in a society in which racism, sexism and poverty are not endemic can all women live full lives. We shall end with a poem by former Ann Arbor City Councilwoman Kathy Kozachenko describing both the treatment women receive when they try to change their way of being-in-the-world and their ultimate triumph.

### Mid-Point

She stored up the anger
for twenty-five years,
then she laid it on the table
like a casserole for dinner.

"I have stolen back
my life," she said.
"I have taken possession
of the rain and the sun
and the grasses," she said.

"You are talking
like a madwoman,"
he said.

"My hands are rocks,
my teeth are bullets,"
she said.

"You are
my wife,"
he said.

"My throat is an eagle
My breasts
are two white hurricanes," she said.

"Stop!" he said.
"Stop or I shall call
a doctor."

"My hair
is a hornet's nest,
my lips
are thin snakes
waiting for their victims."

He cooked his own dinners,
after that.

The doctors diagnosed it
common change-of-life.

She, too, diagnosed
it change of life.
And on leaving the hospital
she said to her woman-friend
"My cheeks
are the wings
of a young
virgin dove.
Kiss them."

## *Notes*

1. We would like to thank the Boston Women's Health Book Collective and Women in Midstream for generously sharing their data with us.
2. Ursula LeGuin, "The Space Crone," *The CoEvolution Quarterly,* Summer 1976, p. 110.
3. Quoted in Carroll Smith-Rosenberg and Charles Rosenberg, "The Female Animal: Medical and Biological Views of Woman and her Role in Nineteenth-Century America," *The Journal of American History,* 60 (1973), 332–356.
4. *Ibid.*
5. Peter Stearns, "Interpreting the Medical Literature on Aging," *Newberry Library Family and Community History Colloquia: The Physician and Social History,* 30 October 1975.
6. John S. Haller, Jr. and Robin M. Haller, *The Physician and Sexuality in Victorian America* (Urbana: University of Illinois Press, 1974) p. 135.
7. *Ibid.,* p. 135.
8. Diana Scully and Pauline B. Bart, "A Funny Thing Happened on the Way to the Orifice: Women in Gynecology Textbooks," *American Journal of Sociology,* 78 (1973), 1045–1050.
9. Margaret A. Campbell, *"Why Would a Girl Go Into Medicine?"* (Old Westbury, N.Y.: The Feminist Press, 1973).
10. Howard W. Jones, Jr., E. J. Cohen, and Robrt B. Wilson, "Clinical Aspects of the Menopause," *Menopause and Aging,* John K. Ryan and D. C. Gibson, eds. (Washington, D.C.: U.S. Government Printing Office, 1971), p. 3.
11. Contrast this view with the feminist perspective offered by June Arnold in the novel *Sister Gin* (Plainfield, Vt.: Daughters, 1975), p. 189: "Bettina will be all right when she reaches menopause. She will be old again as soon as her body stops being under the moon's dominion. The child and the old don't go by clocks and don't know fear. Time took away the child and only time can give her back."
12. Mary Brown Parlee, "Psychological Aspects of the Climacteric in Women" (Paper delivered to the Eastern Psychological Association, New York, April 1976).
13. This is from a drug ad quoted in the Boston Women's Health Book Collective, *Our Bodies, Ourselves* (New York: Simon and Schuster, 1976), p. 327. More recently, under pressure from the Food and Drug Administration, the drug ads are markedly more responsible. There is no evidence, however, that the physicians' attitudes have changed.
14. Robert A. Wilson, *Feminine Forever* (New York: M. Evans and Co., 1966), p. 18.
15. In testimony before the Senate Health Subcommittee, January 21,

1976, then FDA Commissioner Alexander Schmidt said, "It is clear that such treatment [estrogen] must be viewed as providing symptomatic relief for an annoying problem and not as essential therapy for a disease, and that all such treatment is a phenomenon largely of modern Western medicine and medical affluence"—and physicians' desire to keep it that way we might add!

16. See Datha Brack's article in this collection for further discussion.

17. Anita Johnson, "The Risks of Sex Hormones as Drugs," *Women and Health*, 2:1 (1977), pp. 8–11.

18. S. B. Gusberg, "Precursors of Corpus Carcinoma Estrogens and Adenomatous Hyperplasia," *American Journal of Obstetrics and Gynecology*, 54:6 (1947), pp. 905–27.

19. W. V. Gardner, "Carcinoma of the Uterine Cervix and Upper Vagina; Induction under Experimental Conditions in Mice," *Annals of the New York Academy of Science*, 75 (1959), pp. 543–64.

20. Harry K. Ziel and William D. Finkle, "Increased Risk of Endometrial Carcinoma Among Users of Conjugated Estrogens," *New England Journal of Medicine*, 293:23 (1975), pp. 1167–70.

21. Donald C. Smith, Ross Prentice, Donovan J. Thompson, and Walter L. Herrmann, "Association of Exogenous Estrogen and Endometrial Carcinoma," *New England Journal of Medicine*, 293:23 (1975), pp. 1164–7.

22. Robert Hoover, Laman A. Gray, Philip Cole, and Brian MacMahon, "Menopausal Estrogen and Breast Cancer," *New England Journal of Medicine*, 295:8 (1976), p. 501.

23. Sharon Lieberman, "But You'll Make Such a Feminine Corpse ... ," *Majority Report*, 3–4, 19 February-4 March, 1977.

24. "New Discovery: Public Relations Cures Cancer," *Majority Report*, 9–10, 8–18 February, 1977. This letter came to light because an employee in one of the firms felt it important that women know about it, and made the letter available to this feminist newspaper in New York.

25. Morton Mintz, "FDA Requiring Direct Warning to Users of Estrogen," *Washington Post*, 21 July 1977.

26. *F-D-C Reports*, 10–12, 5 September 1977, p. 12 (A newsletter published by the U.S. Food and Drug Administration).

27. And that also was accomplished only by organized pressure from feminist groups.

28. Anita Johnson ("The Risks ... ") says that the suspicion dates back to the 1890s but does not indicate her source.

29. A spot check of physicians after publication of the 1975 studies showed that they had not changed their practices with regard to estrogen prescription. One San Francisco physician compared menopause to diabetes and went on to say: "Most women suffer some symptoms whether they are aware of them or not, so I prescribe estrogens

for virtually all menopausal women for an indefinite period" (*New York Times,* 5 December 1975). A symptom is an outward sign of an underlying problem. That this physician chooses to give carcinogens to women to deal with problems they are not experiencing is the ultimate in medical arrogance and represents medicine's attempt to take over not only our bodies but even our sense of reality. Medical sociologist John McKinlay also has reported that physicians are still prescribing estrogens as frequently as before (10 June 1977).

30. Sonja M. McKinlay and John B. McKinlay, "Selected Studies of the Menopause: An Annotated Bibliography," *Journal of Biosocial Science,* 5 (1973), pp. 533–55.

31. Similarly, one of us (Pauline Bart) found that whether an anthropological study included information about menopause in the culture being studied depended upon whether any of the anthropologists doing the study were women!

32. Sonja M. McKinley and Margot Jeffreys, "The Menopausal Syndrome," *British Journal of Preventive and Social Medicine,* 28:2 (1974), pp. 108–115.

33. "Menopausal" means in the first twelve months following final cessation of menses.

34. Jane E. Prather and Linda S. Fidell, "Sex Differences in the Content and Style of Medical Advertising," *Social Science and Medicine,* 9 (1975), pp. 23–26. Robert Seidenberg, "Drug Advertising and Perception of Mental Illness," *Mental Hygiene,* 55 (1971), pp. 21–31.

35. Paula Weideger, *Menstruation and Menopause* (Revised and Expanded) (New York: Dell, 1977), p. 235.

36. Helene Deutsch, *The Psychology of Women, Volume II* (New York: Grune and Stratton, 1945), p. 459.

37. Howard J. Osofsky and Robert Seidenberg, "Is Female Menopausal Depression Inevitable?" *American Journal of Obstetrics and Gynecology,* 36 (1970), pp. 611–15.

38. Therese Benedek, "Climacterium: A Developmental Phase," *Psychoanalytic Quarterly,* 19:1 (1950), pp. 1–27.

39. Therese Benedek and Boris B. Rubenstein, "Psychosexual Functions in Women," *Psychosomatic Medicine* (New York: Ronald Press, 1952).

40. For critiques and alternative perspectives, *see* Randi Daimon Koeske, "Premenstrual Emotionality: Is Biology Destiny?" *Women and Health,* 1:3 (1976), pp. 11–14; Mary Brown Parlee, "The Premenstrual Syndrome," *Psychological Bulletin,* 80:6 (1973), pp. 454–65; and K. Jean Lennane and R. John Lennane, "Alleged Psychogenic Disorders in Women—A Possible Manifestation of Sexual Prejudice," *New England Journal of Medicine,* 288 (1973), pp. 288–92.

41. Bernice L. Neugarten, Vivian Wood, Ruth J. Kraines, and B. Loomis,

"Women's Attitudes Toward Menopause," in *Middle Age and Aging*, Bernice L. Neugarten, ed. (Chicago: The University of Chicago Press, 1968).

42. The term comes from the Chinese revolutionary groups which gathered to talk about the bad old days in order to purge themselves of the feudal mentality so that they could create a new society. Women's consciousness-raising groups function similarly to help women free themselves from their constricting role socialization.

43. Vivian Wood, one of the authors of that article, has since personally expressed doubt about the reliability of her findings due to the change in mores.

44. Lenore Levit, *Anxiety and the Menopause: A Study of Normal Women* (Doctoral Dissertation, University of Chicago, 1963).

45. Ruth J. Kraines, *The Menopause and Evaluations of the Self: A Study of Middle-aged Women* (Doctoral Dissertation, University of Chicago, 1963).

46. Nancy Dowty, "To Be a Woman in Israel" *School Review*, 80 (1972), pp. 319–332.

47. Pauline B. Bart, "Why Women's Status Changes in Middle Age," *Sociological Symposium*, 3 (1969), pp. 1–18.

48. Cormack cited in Pauline B. Bart, "Why Women's Status Changes . . . ," p. 13.

49. Pauline B. Bart, *Depression in Middle-Aged Women: Some Socio-cultural Factors* (Doctoral Dissertation, University of California at Los Angeles, 1967). *Dissertation Abstracts* 28:4752-B, (University Microfilms No. 68-7452), 1968; see also Bart, "Mother Portnoy's Complaints," *Trans-Action*, 8:1–2 (1970), pp. 69–74; and "Depression in Middle-Aged Women," *Women in Sexist Society*, Vivian Gornick and B. K. Moran, eds. (New York: Basic Books, 1971).

50. Inge P. Bell, "The Double Standard," *Trans-Action*, 8:1–2 (1970), pp. 76–81.

51. Zoe Moss, "It Hurts to be Alive and Obsolete: The Aging Woman," *Sisterhood is Powerful*, Robin Morgan, ed. (New York: Vintage, 1970).

52. Alice Rossi, "Family Development in a Changing World," *American Journal of Psychiatry*, 128 (1972), pp. 1057–80.

53. Bernice L. Neugarten and Nancy Datan, "The Middle Years," *American Handbook of Psychiatry* (2nd ed.), Vol. 1, Salvator Arieti, ed. (New York: Basic Books, 1974).

54. Available from *Prime Time*, 168, W. 86 Street, New York, New York 10024.

55. Another interesting function these groups have performed is the treatment of agoraphobia (fear of open spaces), a widespread problem in Holland. They seem to have succeeded in a number of cases in which years of professional treatment have failed. It is interest-

ing to note that some clinicians regard agoraphobia as the attempt of a person who considers herself powerless to seize some kind of power. Apparently these groups have been able to supply their members with another, more direct, sort of personal power.

[Marlyn Grossman *is a psychologist in Chicago where she was a founding member of Women in Crisis Can Act (W.I.C.C.A.), a feminist hot line, and the Chicago Abused Woman Coalition. While she continues to work actively with both these groups, she is presently working with Pauline Bart on a project investigating women who were attacked and avoided being raped. She does therapy with women and children from a feminist perspective and is a member of the Association for Women in Psychology.*]

[Pauline Bart *is a radical feminist sociologist at the University of Illinois Medical Center. After 12 years of being a trapped suburban housewife, she returned to U.C.L.A. to write her dissertation on depression in middle-aged women (better known as "Portnoy's Mother's Complaint"). While teaching in the usual marginal positions reserved for women at U.S.C. and Berkeley, she presented their first course on women in the Spring of 1969. Most of her work has focussed on the interface between sex roles and health issues— including, with Diana Scully, "A Funny Thing Happened on the Way to the Orifice: Women in Gynecology Textbooks," and with Linda Frankel,* The Student Sociologist's Handbook. *She is a founding mother and active participant in Sociologists for Women in Society, and co-founder and first chair of the Section on the Sociology of Sex Roles of the American Sociological Association. She is currently studying both rape victims and women who successfully avoided rape so that women may be given data-based advice.*]

*Figure XIX.*
*(M.I.T. Historical Collections)*

# Naomi Weisstein

## Adventures of a Woman in Science

I am an experimental psychologist, doing research in vision. The profession has for a long time considered this activity, on the part of one of my sex, to be an outrageous violation of the social order and against all the laws of nature. Yet at the time I entered graduate school in the early sixties, I was unaware of this. I was remarkably naive. Stupid, you might say. Anybody can be president, no? So, anybody can be a scientist. Weisstein in Wonderland. I had to discover that what I wanted to do constituted unseemly social deviance. It was a discovery I was not prepared for: Weisstein is dragged, kicking and screaming, out of Wonderland and into Plunderland. Or Blunderland, at the very least.

What made me want to become a scientist in the first place? The trouble may have started with *Microbe Hunters*,[1] de Kruif's book about the early bacteriologists. I remember reading about Leeuwenhoek's discovery of organisms too small to be seen with the naked eye. When he told the Royal Society about this, most of them thought he was crazy. He told them he wasn't. The "wretched beasties" were there, he insisted; one could see them unmistakably through the lenses he had so carefully made. It was very important to me that he could reply that he had his evidence: evidence became a hero of mine.

It may have been then that I decided that *I* was going to become a scientist, too. I was going to explore the world and discover its wonders. I was going to understand the brain in a better and more complete way than it had been understood before. If anyone questioned me, I would have my evidence. Evidence and reason: my heroes and my guides. I might add that my sense of ecstatic exploration when reading *Microbe Hunters* has never left me through all the years I have struggled to be a scientist.

As I mentioned, I was not prepared for the discovery that women were not welcome in science, primarily because nobody had told me. In fact, I was supported in thinking—even encouraged to think—that my aspirations were perfectly legitimate. I graduated from the Bronx High School of Science in New York City where gender did not enter into intellectual pursuits; the place was a nightmare for everybody.[2] We were all, boys and girls alike, equal contestants; all of us were competing for that thousandth of a percentage point in our grade average that would allow entry into one of those high-class, out-of-town schools, where we could go, get smart, and lose our New York accents.

I ended up at Wellesley and this further retarded my discovery that women were supposed to be stupid and incompetent: the women faculty members at Wellesley were brilliant. I later learned that they were at Wellesley because the schools that had graduated them—the "very best" schools where you were taught to do the "very best" research—couldn't or did not care to place them in similar schools where they could continue their research. So they are our brilliant unknowns: unable to do research because they labor under enormous teaching loads; unable to obtain the minimal support necessary for scholarship; unable to obtain the foundations for productive research: graduate students, facilities, communication with colleagues.

While I was still ignorant about the lot of women in the academy, others at Wellesley were not. Deans from an earlier, more conscious, feminist era would tell me that I was lucky to be at a women's college where I could discover what I was good at and do it. They told me that women in a man's world were in for a rough time. They told me to watch out when I went on to graduate school. They said that men would not like my competing with them. I did not listen to the deans, however; or, when I did listen, I thought what they were telling me might have been true in the nineteenth century, but not in the late fifties.

So my discovery that women were not welcome in psychology began when I arrived at Harvard, on the first day of class. That day, the entering graduate students had been invited to lunch with one of the star professors in the department. After lunch, he leaned back in his chair, lit his pipe, began to puff, and announced: "Women don't belong in graduate school."

The male graduate students, as if by prearranged signal, then leaned back in their chairs, puffed on their newly bought pipes, nodded, and assented: "Yeah."

"Yeah," said the male graduate students. "No man is going to want you. No man wants a woman who is more intelligent than he is. Of course, that's not a real possibility, but just in case. You are out of your *natural* roles; you are no longer feminine."

My mouth dropped open, and my big blue eyes (they have since changed back to brown) went wide as saucers. An initiation ceremony,

I thought. Very funny. Tomorrow, for sure, the male graduate students will get theirs.

But the male graduate students never were told that they didn't belong. They rapidly became trusted junior partners in the great research firms at Harvard. They were carefully nurtured, groomed, and run. Before long, they would take up the white man's burden and expand the empire. But for me and the other women in my class it was different. We were shut out of these plans; we were *shown* we didn't belong. For instance, even though I was first in my class, when I wanted to do my dissertation research, I couldn't get access to the necessary equipment. The excuse was that I might break the equipment. This was certainly true. The equipment was eminently breakable. The male graduate students working with it broke it every week; I didn't expect to be any different.

I was determined to collect my data. Indeed, I *had* to collect my data. (Leeuwenhoek had his lenses. Weisstein would get her data.) I had to see how the experiment I proposed would turn out. If Harvard wouldn't let me use its equipment, maybe Yale would. I moved to New Haven, collected my data at Yale, returned to Harvard, and was awarded my Ph.D. in 1964. Afterward, I could not get an academic job. I had graduated Phi Beta Kappa from Wellesley; had obtained my Ph.D. in psychology at Harvard in two and one half years, ranked first in my graduate class, and I couldn't get a job. Yet most universities were expanding in 1964, and jobs were everywhere. But at the places where I was being considered for jobs, they were asking me questions like, "How can a little girl like you teach a great big class of men?" At that time, still unaware of how serious the situation was, I replied, "Beats me. I guess I must have talent." At another school, a famous faculty liberal challenged me with, "Who did your research for you?" He then put what I assume was a fatherly hand on my knee, and said in a tone of deep concern, "You ought to get married."

Meanwhile, I was hanging on by a National Science Foundation postdoctoral fellowship in mathematical biology at the University of Chicago, attempting to do some research. Prior to my second postdoctoral year, the University of Chicago began negotiations with me for something like a real job: an instructorship jointly in the undergraduate college and the psychology department. The negotiations appeared to be proceeding in good faith, so I wrote to Washington and informed them that I would not be taking my second postdoctoral year. Then, ten days before classes began, when that option as well as any others I might have taken had been closed, the person responsible for the negotiations called to tell me that, because of a nepotism rule—my husband taught history at the University of Chicago—I would not be hired as a regular faculty member. If I wanted to, I could be appointed lecturer, teaching general education courses in the college; there was no possibility of an appointment in psychology. The lectureship paid very little for a lot of work, and I would be teaching material unconnected with my research. Furthermore, a university rule stipulated that lecturers (because their position in the university was so insecure) could not apply for research grants. He concluded by asking me whether I was willing to take the job: ten days before the beginning of classes, he asked me whether I was willing to take the only option still available to me.

I took the job, and "sat in," so to speak, in the office of another dean, until he waived the restriction on applying for research grants. Acknowledging my presence, he told a colleague: "This is Naomi Weisstein. She hates men."

I had simply been telling him that women are considered unproductive precisely because universities do their best to keep women unproductive through such procedures as the selective application of the nepotism rule. I had also asked him whether I could read through the provisions of the rule. He replied that the nepotism rule was informal, not a written statute—flexibility being necessary in its application. Later, a nepotism committee, set up partly in response to my protest, agreed that the rule should stay precisely as it was; that it was a good idea, should not be written out, and should be applied selectively.

Lecturers at major universities are generally women. They are generally married to men who teach at these major universities. And they generally labor under conditions which seem almost designed to show them that they don't belong. In many places, they are not granted faculty library privileges; in my case, I had to get a note from the secretary each time I wanted to take a book out for an extended period. Lecturers' classrooms are continually changed; at least once a month, I would go to my assigned classroom only to find a note pinned to the door instructing me and my class to go elsewhere: down the hall, across the campus, out to Gary, Indiana.

In the winter of my first year, notices were distributed to all those teaching the courses I was teaching, announcing a meeting to discuss the following year's syllabus. I didn't receive the notice. As I was to learn shortly, this is the customary way a profession that prides itself on its civility and genteel traditions indicates to lecturers and other "nuisance personnel" that they're fired: they are simply not informed about what's going on. I inquired further. Yes, my research and teaching had been "evaluated" (after five months: surely enough time), and they had decided to "let me go" (a brilliant euphemism). Of course, the decision had nothing to do with my questioning the nepotism rules and explaining to deans why women are thought unproductive.

I convinced them to "let me stay" another year. I don't know to this day why they changed their minds. Perhaps they changed their minds because it looked like I was going to receive the research grant for which I had applied, bringing in money not only for me, but for the university as well. A little while later, Loyola University in Chicago offered me a job.

So I left the University of Chicago. I was awarded the research grant and found the Psychology Department at Loyola at first very supportive. The chairman, Ron Walker, was especially helpful and enlightened about women at a time when few academic men were. I was on my way, right? Not exactly. There is a big difference between a place like Loyola and a place with a heavy commitment to research—a large state university, for example—a difference that no amount of good will on the part of an individual chairman can cancel out. The Psychology Department was one of the few active departments at Loyola. The other kinds of support one needs to do experimental psychology—machine and electrical shops, physics and electrical engineering departments, technicians, a large computer—were either not available or were available at that time only in primitive form.

When you are a woman at an "unknown" place, you are considered out of the running. It was hard for me to keep my career from "shriveling like a raisin" (as an erstwhile colleague predicted it would). I was completely isolated. I did not have access to the normal channels of communication, debate, and exchange in the profession—those informal networks where you get the news, the comment and the criticism, the latest reports of what is going on. I sent my manuscripts to various people for comment and criticism before sending them off to journals; few replied. I asked others working in my field to send me their prepublication drafts; even fewer responded. Nobody outside Loyola informed me about special meetings in my area of psychology, and few inside Loyola knew about them. Given the snobbery rife in academic circles (which has eased lately since jobs are much harder to find and thus even "outstanding" young male graduates from the "best" schools may now be found at places formerly beneath their condescension), my being at Loyola almost automatically disqualified me from the serious attention of professional colleagues.

The "inner reaches" of the profession, from which I had been exiled, are not just metaphorical and intangible. For instance, I am aware of two secret societies of experimental psychologists in which fifty or so of the "really excellent" young scientists get together regularly to make themselves better scientists. The ostensible purpose of these societies is to allow these "best and brightest" young psychologists to get together to discuss and criticize each other's work; they also function, of course, to define who is excellent and who is not, and to help those defined as excellent to remain so by providing them with information to which "outsiders" in the profession will not have access until much later (if at all).

But the intangibles are there as well. Women are subjected to treatment men hardly ever experience. Let me give you a stunning example. I wrote up an experiment with results I thought were fascinating, and sent the paper to a journal editor whose interests I knew to be close to what was reported in my paper. The editor replied that there were some control conditions that should be run and some methodological loose ends; so they couldn't publish the paper. Fair enough. He went on to say that they had much better equipment over there, and they would like to test my idea themselves. Would I mind? I wrote back, and told them I thought it was a bit unusual, asked if they were suggesting a collaboration, and concluded by saying that I would be most happy to visit with them and collaborate on my experiment. The editor replied with a nasty letter explaining to me that by suggesting that they test my idea themselves, they had merely been trying to help me. If I didn't want their help in this way, they certainly didn't want mine; that is, they had had no intention of suggesting a collaboration.

In other words, what they meant by "did I mind" was: Did I mind if they took my idea and did the experiment themselves? As we know, taking someone else's idea and pretending it's your own is not at all an uncommon occurrence in science. The striking thing about this exchange was, however, that the editor was arrogant enough, and assumed that I would be submissive enough, for him to openly ask me whether I would agree to this arrangement. Would I mind? No, of course not. Women are joyful altruists. We are happy to give of ourselves. After all, how many good ideas do you get in your lifetime? One? Two? Why not *give* them away?

Generally, the justification for treating women in such disgraceful ways is simply that they are women. Let me give another example. I was promised the use of a small digital laboratory computer, which was to be purchased on a grant. The funds from the grant would become available if a certain job position entailing administration of this grant could be filled. I was part of the group which considered the candidates and which recommended appointing a particular individual. During the discussion of future directions of the individual's work, it was agreed that he would, of course, share the computer with me. He was hired, bought the computer, and refused me access to it. I offered to put in money for peripherals which would make the system faster and easier for both of us to work with, but this didn't sway him. As justification for his conduct, the man confessed to the chairman that he simply couldn't share the computer with me: he had difficulty working with women. To back this up, he indicated that he'd been "burned twice." Although the chairman had previously been very helpful and not bothered in the least about women, he accepted that statement as an explanation. Difficulty in working with women was not a problem this man should work out. It was *my* problem. Colleagues thought no worse of him for this problem; it might even have raised him in their estimation. He obtained tenure quickly, and retains an influential voice in the department. Yet if a woman comes to *any* chairman of *any* department and confesses that she has difficulty working with *men,* she is thought pathological.

What this meant for me at the time was that my research was in jeopardy. There were experimental conditions I needed to run that simply could not be done without a computer. So there I was, doing research with stone-age equipment, trying to get by with wonder-woman reflexes and a flashlight, while a few floors below, my colleague was happily operating "his" computer. It's as if we women are in a totally rigged race. A lot of men are driving souped-up, low-slung racing cars, and we're running as fast as we can in tennis shoes we managed to salvage from a local garage sale.

Perhaps the most painful of the appalling working conditions for women in science is the peculiar kind of social-sexual assault women sustain. Let me illustrate with a letter to *Chemical and Engineering News* from a research chemist named McGauley:

> There are differences between men and women . . . just one of these differences is a decided gap in leadership potential and ability . . . this is no reflection upon intelligence, experience, or sincerity. Evolution made it that way . . . . Then consider the problems that can arise if the potential employee, Dr. Y (a woman) [*sic:* he could at least get his chromosomes straight] will be expected to take an occasional business trip with Dr. X. . . . Could it be that the guys in shipping and receiving will not take too kindly to the lone Miss Y?[3]

Now what is being said here, very simply, and to paraphrase the Bible, is that women are trouble. And by trouble, McGauley means sexual trouble. Moreover, somehow, someway, it is our fault. *We* are provoking the guys in shipping and receiving. Women—no matter who the women are or what they have in mind—are universally assigned by men, first, to sexual categories. Then, women are accused by men of taking their minds away from work. When feminists say that women are treated as sex objects, we are compressing into a single, perhaps rhetorical phrase, an enormous area of discomfort, pain, harassment, and humiliation.

This harassment is especially clear at conventions. Scientific meetings, conferences, and conventions are harassing and humiliating for women because women, by and large, cannot have male colleagues. Conversations, social relations, invitations to lunch, and the like are generally viewed as sexual, not professional, encounters if a woman participates in them. It does not cross many men's minds that a woman's motivation may be entirely professional.

I have been at too many professional meetings where the "joke" slide was a woman's body, dressed or undressed. A woman in a bikini is a favorite with past and perhaps present presidents of psychological associations. Hake showed such a slide in his presidential address to the Midwestern Psychological Association, and Harlow, past president of the American Psychological Association, has a whole set of such slides, which he shows at the various colloquia to which he is invited. This business of making jokes at women's bodies constitutes a primary social-sexual assault. The ensuing raucous laughter expresses the shared understanding of what is assumed to be women's primary function to which we can always be reduced. Showing pictures of nude and sexy women insults us: it puts us in our place. You may think you are a scientist it is saying, but what you really are is an object for our pleasure and amusement. Don't forget it.

I could continue recounting the horrors, as could almost any woman who is in science or who has ever been in science. But I want now to turn to the question of whether or not the commonly held assumptions about women are true. If they are, this would in no way justify the profession's shameful treatment of women, but it might lead us to different conclusions about how to remedy the situation.

I began the inquiry into women's "nature" while I was in graduate school. I wanted to investigate the basis on which the learned men in my field had pronounced me and my female colleagues unfit for graduate study.

I found that the views of the experts reflected, in a surprisingly transparent way, the crudest cultural stereotypes. Erik Erikson wrote:[4]

Young women often ask me whether they can 'have an identity
before they know whom they will marry and for whom they will
make a home'

He explained (somewhat elegiacally) that:

Much of a young woman's identity is already defined in her
kind of attractiveness and in the selectivity of her search for
the man (or men) by whom she wishes to be sought. . . .

Mature womanly fulfillment, for Erikson, rested on this "fact":

[that a woman's] . . . somatic design harbors an 'inner space'
destined to bear the offspring of chosen men, and with it, a
biological, psychological, and ethical commitment to take care
of human infancy.

Bruno Bettelheim, speaking at a symposium on American women in
science and engineering, commented:

We must start with the realization that, much as women want to
be good scientists and/or engineers, they want first and foremost to be womanly companions of men and to be mothers.[5]

And Joseph Rheingold, a psychiatrist at the Harvard Medical School,
tied the reluctance of women to give in to their true natures to society's
problems:

Woman is nurturance . . . anatomy decrees the life of a woman
. . . When women grow up without dread of their biological
functions and without subversion by feminist doctrine, and
therefore enter upon motherhood with a sense of fulfillment and
altruistic sentiment, we shall attain the goal of a good life and
a secure world in which to live it.[6]

So the learned men in my field were saying essentially the same
thing as the male graduate students, but it still did not sound right to
me. Since scientists (supposedly) do not assess the truth or falsity of
propositions on the basis of who said them (they look instead for
evidence), I decided to look for the evidence on which these eminent
men had based their theories. I determined that there is no evidence
behind their fantasies of the servitude and childish dependence of women.
On the contrary, the idea of human possibility which rests on the accident of sex at conception has strangled and deflected psychology so
that it is still relatively useless in describing, explaining, or predicting
human behavior. This is true for men as well as women. It becomes

especially pernicious when the theories are not only wrong but are proscriptive as well.

I have elsewhere gone through the arguments showing the near use-lessness of these kinds of theories of human nature;[7] here, let me just provide the briefest of summaries.

The basic reason that this kind of psychology (i.e., personality theory, and, for the most part, theory from psychotherapists and psychoanalysts) tells us next to nothing about human nature, male as well as female, is that it has been looking in the wrong place. It has assumed that what people do comes from a fixed, rigid, inside directive: sex organs, or fixed cognitive traits, or what happened until, but no later than, the age of 5. This assumption has been shown to fail again and again in tests; a person will be assessed by psychologists as possessing a particular constellation of personality traits and then, when different criteria are applied, or someone different is asked to judge, or, more importantly, when that person is in a different kind of social situation, s/he will be thought to exhibit a completely different set of traits.[8]

One might argue, then, that personality is a somewhat subtle and elusive thing, and that it would be difficult to obtain a set of measures that would distinguish personality types. This is a reasonable argument. But even when one looks at what one would expect to be gross dif-ferences between a certified schizophrenic, say, and a normal,[9] or between a male homosexual and a male heterosexual,[10] or between a "male" personality and a "female" personality,[11] one finds that the same judges who claim to be able to differentiate human personalities, simply cannot distinguish one from the other. In one study,[12] for ex-ample, judges who were supposed to be experts at this kind of thing could not tell, on the basis of clinical tests and interviews (in which one is allowed to ask questions such as, "When did you first notice that you had grown antlers?"), which one of a group of people had been classified as schizophrenics and which as normals. Even stranger, some weeks later, when the same judges were asked to judge the same peo-ple, in many cases, they reversed their own judgments. Judges (again, allegedly clinical experts), attempting to distinguish between homo-sexuals and heterosexuals on the basis of what is assumed to be dif-ferences in their personalities, have done no better; nor was my grad-uate class at Harvard able correctly to distinguish stories written by males from those written by females, even though we had just com-pleted a month and a half's study of the differences between men and women. In short, if judges cannot agree on whether a person belongs in a certain personality category, even when those categories are assumed to be as different from each other as normal/crazy, male/female, and straight/gay; if the measurement depends on who is doing the measuring, and on what time of day it is being done, then theories that are based on these personality categories are useless.

The other "test" that has frequently been cited as a way of confirming such theories is the test of therapy. Since most of the theorizing about men and women (and normals and schizophrenics, and homosexuals and heterosexuals) has been done by clinical psychologists who cite as evidence for their theories "years of intensive clinical experience," one test of their understanding of human personality might be their effectiveness in helping people solve these "problems." Of course, one might question what is going on in these years of intensive clinical practice when clinicians cannot even agree on descriptions; that is, they cannot agree on their categories. But suppose one countered that clinical psychologists really do have an understanding of the depths of human personality, and, although any two clinical psychologists may not be able to agree on a "verbal" level on what categories they are using, nevertheless (so this argument would go), they are operating at an intuitive level which "works" (i.e., they help their patients change their behavior). The fact is that, to the limited extent that therapy may change behavior (if at all), it doesn't matter *which* therapy is used: in general, no one therapy is reported to be any *more* effective than any other, even when the same symptoms are being treated.[13] Since theories upon which different therapies are based are different, and in some cases, conflict with each other, the extent to which a particular therapy may work cannot be taken to lend credence to that particular theory.

What all this means for women is that personality theory has given us no idea of what our true "natures" are; whether we were intended from the start to be scientists and engineers and were thwarted by a society that has other plans for us, or whether we were intended, as claimed by some of the learned men in the field, only to be mothers. There are a number of arguments based on selected primates[14] that also purport to show that females are suited only for motherhood (hopefully with a sense of fulfillment and altruistic sentiment). These arguments are even more specious than those from the clinical tradition, as I have discussed elsewhere.[15]

But while personality psychology and clinical psychologists have failed miserably at providing any statements we can trust about women's "true nature," or about anyone's "true nature," the evidence is accumulating from a different area of psychology, *social* psychology: what humans do and when they will do it is highly predictable. What people do and who they believe themselves to be will, in general, be a function of what the people around them expect them to be, and what the overall situation in which they are acting implies that they are. Let me describe three experiments that have made this fact clear.

## The Experimenter Bias Experiments

These studies[16] have shown that if one group of experimenters has one hypothesis about what they expect to find, and another group of experimenters has the opposite hypothesis, both will obtain results in accord with their differing hypotheses. And this is not because the experimenters lie or cheat or falsify data. In the studies cited, the experimenters are closely observed, and they are made outwardly to behave in identical fashion. The message about their different expectations is somehow picked up by their subjects through nonverbal cues, head nods, ways of communicating expectations that we do not yet know about. The moral here is that, even in carefully controlled conditions, when we are dealing with humans (and in some cases rats),[17] the hypotheses we start with will influence the behavior of the organism we are studying. It is obvious how important this would be when assessing the validity of psychological studies of the differences between men and women.

## Inner Physiological State Versus Social Context

Subjects[18] were injected with adrenalin, a hormone that tends to make people "speedy"; when placed in a room with another person (a confederate of the experimenter) who acted euphoric, the subject became euphoric. Conversely, if a subject was placed in a room with another person who acted angry, the subject became angry. These data seem to indicate that the far more important determinant to how people will act is not their physiological state, but the social context in which they are acting. Thus, no matter how many physiological differences we may find between men and women, we must be very cautious in assigning any fixed behavioral correlates to the physiological states. The point is made even more strongly, perhaps, in studies of hermaphrodites in whom the genetic, gonadal, hormonal sex, the internal reproductive organs, and the ambiguous appearance of the external genitalia were identical. It was shown that one will consider one's self male or female depending simply on whether one was defined and raised as a male or a female:

There is no more convincing evidence of the power of social interaction on gender-identity differentiation than in the case of congenital hermaphrodites who are of the same diagnosis and similar degree of hermaphroditism but are differently assigned and with a different postnatal medical and life history.[19]

## The Obedience Experiments

In Milgram's experiments,[20] a subject is told that s/he is administering a learning experiment, and that s/he is to deal out shocks each time another "subject" (who is in fact a confederate of the experimenter) answers incorrectly. The equipment appears to provide graduated shocks ranging upwards from 15 V through 450 V; for each of four consecutive voltages, there are verbal descriptions such as mild shock; danger; severe shock; and finally, for the 435 V and 450 V switches a red XXX marked over the switches. Each time the confederate answers incorrectly, the subject is supposed to increase the voltage. As the voltage increases, the confederate begins to cry out in pain and demands that the experiment be stopped, finally refusing to answer at all. When all responses are stopped, the experimenter instructs the subject to continue increasing the voltage. For each shock administered, the confederate shrieks in agony. Under these conditions, about 62 per cent of the subjects administered shocks that they believed to be possibly lethal.

No tested individual differences among subjects predicted who would continue to obey, and who would break off the experiment. When forty psychiatrists predicted how many of a group of 100 subjects would go on to give the lethal shock, their predictions were orders of magnitude below the actual percentages; most expected only one or two of the subjects to obey to the end.

But even though psychiatrists have no idea how people will behave in this situation, and even though individual differences do not predict who will and will not obey, it is easy to predict when subjects will be obedient and when they will be defiant. In a variant of Milgram's experiment, two confederates were present in addition to the "victim," working along with the subject in administering electric shocks. When the two confederates refused to continue with the experiment, only 10 per cent of the subjects continued to the maximum voltage. This is critical for personality theory. It says that behavior is predicated largely on the social situation, not solely on the individual's history.

To summarize: if subjects under quite innocuous and noncoercive social conditions can be made to kill other subjects, and under other types of social conditions will positively refuse to do so; if subjects can react to a state of physiological arousal by becoming euphoric because there is someone else around who is euphoric, or angry because there is someone else around who is angry; if subjects will act a certain way because experimenters expect them to act in that way, and another group of subjects will act in a different way because experimenters expect them to act in that different way; then it appears obvious that a study of human behavior requires first and foremost a study of the social contexts within which people move, the expectations as to how they will behave, and the authority that tells them who they are and what they are supposed to do.

The relevance to males and females is obvious. We do not know what immutable differences in behavior, nature, ability, or possibility exist between men and women. We know that they have different genitalia and at different times in their lives, different sex hormone levels. Perhaps there are some unchangeable differences; probably there are a number of irrelevant differences. But all these differences are likely to be trivial compared to the enormous influence of social context. And it is clear that, until social expectations for men and women are equal and just; until equal respect is provided for both men and women, our answers to the question of immutable differences, of "true" nature, of who should be the scientist and who should be the secretary, will simply reflect our prejudices.

I want to stop now and ask: What conclusions can we draw from my experience? What does it all add up to?

Perhaps we should conclude that persistence wins out. Or that life is hard, but cheerful struggle and a "sense of humor" can make it bearable. Or perhaps we should search back through my family, and find my domineering mother and passive father, or my domineering father and passive mother, to explain my persistence. Perhaps . . . but all these conclusions are beside the point. The point is that none of us should have to face this kind of offense. The main point is that we must change this man's world and this man's science.

How will other women do better? One of the dangers of this kind of narrative is that it may validate the punishment as it singles out the few survivors. The lesson appears to be that those (and only those) with extraordinary strength will survive. This is not the way I see it. Many have had extraordinary strength and have *not* survived. We know of some of them, but by definition we will of course never know of most.

Much of the explanation for my own professional survival has to do with the emergence and growth of the women's movement. I am an experimental psychologist, a scientist. I am also a feminist. I am a feminist because I have seen my life and the lives of women I know harassed, dismissed, damaged, destroyed. I am a feminist because without others I can do little to stop the outrage. Without a political and social movement of which I am a part, without feminism, my determination and persistence, my clever retorts, my hours of patient explanation, my years of exhortation amount to little. If the scientific world has changed since I entered it, it is not because I managed to become an established psychologist within it. Rather, it is because a women's movement came along to change its character. It is true that as a member of that movement, I have acted to change the character of the scientific world. But without the movement, none of my actions would have brought about change. And now, as the strength of the women's movement ebbs, the old horrors are returning. This must not happen.

Science, knowledge, the search for fundamental understanding is part of our humanity. It is an endeavor that seems to give us some glimpse of what we might be and what we might do in a better world. To deny us the right to be scientists is to deny us our humanity. We cannot let that happen.

## Notes

Somewhat different versions of this paper have appeared in *Federation Proceedings*, 35 (1976), 2226–31; and in *Working It Out,* S. Ruddick and P. Daniels, eds., (New York: Pantheon Books, 1977), pp. 242–250. I wish to thank Tobey Klass for her critical comments and her guidance through the recent literature. I also wish to thank Roger Burton for bringing the Campbell and Yarrow sudy to my attention.

1. Paul de Kruif, *Microbe Hunters* (New York: Brace & World, 1926).
2. I discovered later on that this in itself was unusual—by high school, if not before, girls are generally discouraged from showing an interest in science.
3. P. J. McGauley, Letter to the Editor, *Chem. Eng. News, 48* (1970), pp. 8–9.
4. Erik H. Erikson, "Inner and Outer Space: Reflections on Womanhood," *Daedalus,* 93 (1964), pp. 585–606.
5. From a speech entitled "The Commitment Required of a Woman Entering a Scientific Profession in Present Day American Society," Massachusetts Institute of Technology, 1965.
6. J. Rheingold, *The Fear of Being a Woman* (New York: Grune & Stratton, 1964).
7. Naomi Weisstein, "Psychology Constructs the Female; or the Fantasy Life of the Male Psychologist," *J. Soc. Ed.,* 35 (1970), pp. 362–373.
8. J. Block, "Some Reasons for the Apparent Inconsistency of Personality," *Psychol. Bull.,* 70 (1968), 210–212; W. Mischel, *Personality and Assessment* (New York: Wiley, 1968); W. Mischel, "Toward a Cognitive Social Learning Reconceptualization of Personality," *Psychological Review,* 80 (1973), pp. 252–283.
9. K. B. Little and E. S. Schneidman "Congruences Among Interpretations of Psychological and Anamnestic Data," *Psychol. Monogr.,* 73 (1959), pp. 1–42. See also R. E. Tarter, D. I. Templer, and C. Hardy, "Reliability of the Psychiatric Diagnosis," *Diseases of the Nervous System,* 36 (1975), pp. 30–31.
10. E. Hooker, "Male Homosexuality in the Rorschach," *J. Projective Techniques,* 21 (1957), pp. 18–31.
11. Naomi Weisstein, "Psychology Constructs the Female . . . ."
12. K. B. Little and E. S. Schneidman, "Congruences among interpretations . . . ."
13. Dorothy Tennov, *Psychotherapy* (New York: Abelard-Schuman, 1975); M. L. Smith and G. V. Glass, "Meta-Analysis of Psychotherapy Outcome Studies," *American Psychologist,* 32 (1977), pp. 752–60; Sloan, Staples, Cristol, Yorkston and Whipple, "Short-term Analytically Oriented Psychotherapy Versus Behavior Therapy," *American Journal of Psychiatry,* 132:4 (1975), pp. 373–377.

14. See H. F. Harlow, "The Heterosexual Affectional System in Monkeys," *Am. Psychol.*, 17 (1962), pp. 1–9; L. Tiger, *Men in Groups* (New York: Random House, 1969); and L. Tiger, "Male Dominance? Yes. Alas. A Sexist Plot? No.," *New York Times Magazine Section*, 25 October 1970.

15. Naomi Weistein, "Psychology Constructs the Female . . . ." *See also* Hubbard's article in this collection.

16. R. Rosenthal, "On the Social Psychology of the Psychological Experiment: the Experimenter's Hypothesis as Unintended Determinant of Experimental Results," *Am. Sci.*, 51 (1963), 268–283; and R. Rosenthal, *Experimenter Effects in Behavioral Research* (New York: Appleton-Century-Crofts, 1966). A meticulous observation of behavior at a summer camp suggesting the same results is J. D. Campbell and M. R. Yarrow's "Perceptual and Behavioral Correlates of Social Effectiveness," *Sociometry*, 24 (1961), pp. 1–20.

17. H. F. Harlow, "The Heterosexual Affectional System . . . ."

18. John Money, "Sexual Dimorphism and Homosexual Gender Identity," *Psychol. Bull.*, 74 (1970), pp. 6, 425–440; and S. Schachter and J. E. Singer, "Cognitive, Social, and Physiological Determinants of Emotional State," *Psychol. Rev.*, 63 (1962), pp. 379–399.

19. John Money, "Sexual Dimorphism and Homosexual Gender . . . ."

20. S. Milgram, "Some Conditions of Obedience and Disobedience to Authority," *Human Relations*, 18 (1965), pp. 57–76; and S. Milgram, "Liberating Effects of Group Pressure," *J. Pers. Soc. Psychol.*, 1 (1965), pp. 127–134.

[Naomi Weisstein *is Professor of Psychology at SUNY at Buffalo. Her research is in vision, perception, cognition and brain theory, focusing on how images are represented in the visual system so as to lead to human visual understanding. She is a member of many professional organizations and has published numerous scientific papers and reviews. During the 1960's, she was active in CORE and formed women's caucuses in SDS and the New University Conference. In the early women's liberation movement, she was a founding member of Chicago West Side Group and the Chicago Women's Liberation Union. She has been active in guiding and defining an insurgent feminist culture, as organizer and pianist for the Chicago Women's Liberation Rock Band (1970–73), writer and critic, and currently, as feminist comedienne presenting themes ranging from rape, to the mores of the scientific pro-. fession, to attempts to deliver scientific data as stand-up monologue.*]

*Figure XX.   Dr. Ellen Swallow Richards, first woman to serve on M.I.T. teaching staff, with group of female students (1888).*
*(M.I.T. Historical Collections)*

# Epilogue

What do women expect from science? What can it do for us, and we for it?

The first thing it can do *for* us is to stop doing a number of unpleasant things *to* us. Or in the words of Sarah Grimke written in 1837 to her sister, Angelina:

> All I ask of our brethren is, that they will take their feet from off our necks and permit us to stand upright on that ground which God designed us to occupy.[1]

We offer the following to our brethren in science as a model for the expeditious removal of feet. Recently, it was brought to the attention of E. B. White that his *Elements of Style,* which has been a writers' bible for twenty years, was outrageously and gratuitously sexist. Among its hundreds of examples, illustrating every disservice we are likely to render the English language, not *one* presents a positive image of women. White, not one to waste words on superfluities, did not waste them here. "I'll fix it," he said according to a July 1977 news item. "I wasn't aware."

If every man who was ever confronted with the anti-female bias of his work—be he biology professor, grant reviewer, researcher, medical school admissions officer, textbook writer—were to adopt the above example as *method,* we could soon redress some past grievances. And we could all get on to the business of building a better future. But that is unfortunately not likely to happen.

The last two or three years have witnessed the rebirth of biological determinism in the guise of the new science of sociobiology. Well-seasoned with scientific sexism, it includes a revived form of Darwin's theory of sexual selection and tries to construct an evolutionary underpinning for human sex role behavior.[2] Nor has the impact of sociobiology been limited to the scientific literature. Sociobiology has become a media event, with popular magazine articles[3] and a film that depicts professors discussing their theories amidst footage of muscular football players, long-legged young women, cavorting baboons, and pregnant African wood carvings.

The work of John Money on sex and gender and the research on brain asymmetry are being used to lend scientific "authority" to new theories of sex differences, more comprehensively androcentric than their predecessors.[4]

It is clear that progress in forcing a redefinition of what women are —or better yet, an end to the effort to define us—is going to be slow and will be met with vigorous efforts to prove again and again that our social disadvantages are grounded in biology. Indeed, "progressives" may even offer us "compensatory" education for our (innate) leadership deficiencies while promising men remedial training for their (congenitally) defective "parenting."[5]

Sometimes it doesn't seem that we have moved very far beyond Darwin's Victorian credo, or the psychologist Bettelheim's restatement of it in 1965:

> We must start with the realization that, much as women want to be good scientists and/or engineers, they want first and foremost to be womanly companions of men and to be mothers.

We have not even won that first critical victory: to have our dissatisfaction with the current state of affairs be other than a source of amusement to those responsible for it. The routinely jocular dismissal of charges of misogyny brings to mind Kate Millett's remark that "sexism (unlike most other systems of oppression) is pleased with itself."

So, to answer the original questions, one of the most liberating things androcentric scientists could do for women is to stop telling us what we are, while a male-dominated society sets the limits to what we can be.[6] Another is to let us control those aspects of science that most directly affect our lives. Yet here, too, a happy ending is not in sight. Women's health care is still delivered primarily by men. Midwifery is still illegal in most parts of the United States, and if the number of states that allow it is increasing, that is probably owing to the increasing expense of hospital-based medical care rather than to a commitment to women-controlled birthing.[7]

While under the watchful eye of Title IX there are now more women in medical and graduate schools than ever before, numbers tell only part of the story. Women are being admitted, but what will become of them after they graduate? It looks as though male doctors may cede the less lucrative and prestigious field of primary care to women (a kind of occupational "there goes the neighborhood") and attempt to retain their dominant position in the more "elite" male bastions of surgery, medical research, and hospital administration. And the same pattern threatens in academic science, where in the shrinking job market, more females are now permitted to enter as instructors, research fellows, or untenured assistant professors. But will more than a few rise to the tenured positions that bring greater freedom to choose what to study and more money with which to do it?

Without this freedom, all other inroads into science will avail women little. For slowly we realize that present research trends often provide us with options we would rather not have and force us to make choices we would rather not make. (Do we want to get bloodclots from the pill or have our uteruses perforated by IUD's? Would we rather burn garbage, which wastes its nutrients while polluting the air, or dump it into the sea?) The increasing pace of science and its applications as technology insure that the future will confront us with more such choices, not fewer.

Unless we participate in the processes that generate the options, when it comes to a final choice—among methods of birth control, childbirth, or abortion; among means of diagnosing, curing, preventing, or caring for illness; among methods of growing food, generating electric power, heating our houses—we will find ourselves always unhappy choosers between lesser and greater evils. And until we develop the expertise needed to create our own options, we will not be able to see beyond those evils, and will believe them to be the actual limits to our choices—which is, of course, the way they are usually portrayed.

Unfortunately, in the last three centuries, science has become a monolith in which only certain ways of viewing nature and learning about it are acceptable. The technology that derives from it is equally monolithic; its main charge is: go forth and multiply profits. For this reason, we are usually presented with a very limited range of options, generated by "experts" and with essentially no public debate. It is not easy for feminists to provide alternatives. Women are still vastly underrepresented in the top ranks of science, where decisions are made about meaning, aims, strategies and tactics; we still cluster at the bottom of the work-force that makes science and technology "run."[8] The few of us who find our way into the classrooms where we can learn to become scientists, doctors, and engineers rarely derive from our isolated and precarious situation the courage to question the system of presuppositions on which present-day science and technology are built. Indeed, we may not even recognize the existence of these questions any more than do our male peers.

Even when it has been our lot (as it often is) to have lived a sufficiently marginal existence in the traditional schools and disciplines so that we tend to question accepted views, we are still in a double bind: if our questions are too heterodox, they suffice to disqualify us as proper "material" for training; but if we suppress such questions sufficiently deep and long, we may stop thinking them and emerge from our educations as the monolith's true devotees. This is likely the reason why few women doctors have been much help to feminists working for health care alternatives. And it explains even more the orthodoxy of female scientists; for while "alternative" doctors have some measure of economic and intellectual independence as entrepreneurs, scientists work almost exclusively within the hierarchical, male-dominated academic-technical establishment where employment, research funds, ability to publish one's findings, and tenure heavily depend upon conformity to the established view.

There is another problem women must face if we decide to effect change by becoming scientists: how to reconcile our lives with the traditional structure of the male-dominated professions. From the models of the old-world monastic scholar and the upper class gentleman-scientist of independent means, has come the bourgeois scientist of today, whose success depends upon having assistants, mostly females (wife, secretary, technicians . . . ), to nourish and nurture *him* in *his* single-minded search for truth.

Women are told that if we want to enter the highly competitive rat-race that is said to be the *only* way to do *real* science, we must adopt the lifestyle and manner of the male researcher. What does that mean? That we must not form egalitarian relationships at home and at work, not marry or have children if we choose to? Obviously that is not enough. Prototypic male scientists not only reject family responsibilities (which is not to say that they don't have families; only that wives and children must know their place in the scientists' hierarchy of commitments and values); they also exploit women's labor, as unpaid wife and as the underpaid "support staff" that makes modern laboratories run.

The social structure of scientific work as well as its products reflect its male origins, and both will have to change before scientific work will be a genuine option for large numbers of women. E. O. Wilson may have been correct when he wrote in the *New York Times Magazine* that "Even with identical education and equal access to all professions, men are likely to play a disproportionate role in political life, business, and science." But this is not, as he would have us believe, because our evolutionary history has endowed women with domestic and nurturing genes and men with professional ones, but because conditions of work in the male-dominated professions do not suit the lives most women want or are able to live.

Another possibility awaiting feminist scientists is to redress the exploitative relationship of androcentric science and technology to nature. Can we come up with ways to think about nature and to probe its workings that do not degrade, kill and destroy it—ways that recognize that we, too, are its creatures? That is the chief methodological challenge that confronts the feminist revulsion against the Baconian warrior who has been riding astride nature like the bombadier-cowboy in Kubrick's *Doctor Strangelove* as he yodels off into space to destroy the world.

The man-nature antithesis was invented by men. Our job is to re-invent a relationship that will realize (in the literal sense of making real) the unity of humankind with nature and will try to understand its workings from the inside—ever cognizant of the fact that our lives are inextricably bound up with it. Science is just one way of making sense of natural events; there are others. It is up to us to select from among them those that yield a sense that is consonant with our ideas of human dignity and the dignity of nature free from human exploitation. We must also insist that the technology we build must improve our lives as *we* define improvement, not as it is defined for us by the "experts."

We have tried in this book to make a beginning at a new science by raising new questions and groping towards new answers. We hope it will impel others to look critically at what they accept as real and significant about the natural world. If we want science and technology to serve us, we must change their present course.

Science is a human construct that came about under a particular set of historical conditions when *men's* domination of nature seemed a positive and worthy goal. The conditions have changed and we know now that the path we are travelling is more likely to destroy nature than to explain or improve it. Women have recognized more often than men that we are part of nature and that its fate is in human hands that have not cared for it well. We must now act on that knowledge.

*October, 1977*

## Notes

1. Letter of July 17, 1837, as reprinted in *Feminism: The Essential Historical Writings,* Miriam Schneir, ed. (New York: Alfred A. Knopf, Bantam Edition, 1972) p. 30.
2. For a discussion of sexism in sociobiology, *see* pages 21–26 in this book. Detailed critiques have also been written by Barbara Chasin, "Sociobiology: A Sexist Synthesis," *Science for the People,* 9 (May-June 1977), pp. 27–31; Freda Salzman, "Are Sex Roles Biologically Determined?" *Science for the People,* 9 (July-Aug. 1977), pp. 37–32; and Evelyn Reed, *Sexism and Science,* New York: Pathfinder Press, 1978.
3. *See,* for example. E. O. Wilson, "Human Decency is Animal," *New York Times Magazine,* 12 Oct. 1975, pp. 38–50; and the cover story, "Why You Do What You Do–Sociobiology: A New Theory of Behavior," *Time,* 1 August 1977.
4. One of the most striking examples of this domino effect appeared in an account of the proceedings of a meeting of the American Association for the Advancement of Science in May, 1977. According to a *Boston Herald American* report, Dr. William Blackmore of Toronto reported "evidence" that the relatively small number of female creative geniuses this world has produced is attributable to innate biological differences between women and men. His "evidence" turns out to be none other than Money's research on rat brains and fetally androgenized girls, as well as a garbled report of the alleged differences in the patterns of hemispheric dominance in the brains of women and men. (*See also* Fried's and Star's essays.)
5. It is interesting that in English, the verbs "to mother" and "to father" so completely reflect the patriarchal social structure in which fathers contribute sperm and mothers childcare that a third verb, "to parent," must be used to describe the activity of fathers caring for their children.
6. "Capabilities," writes Simone de Beauvoir, "are clearly manifested only when they have been realized." *The Second Sex* (New York: Alfred A. Knopf, Bantam Edition, 1952), p. 30.
7. *See* article by Datha Clapper Brack in this book.
8. This work-force is a pyramid, at the bottom of which cluster large numbers of underpaid women *and* men, while at the top are a small number of highly paid, male professionals. For example, in the health services in the U.S. in 1970, 71.9% of employees were women, but 98.2% of them worked as therapists, nurses and other

paraprofessional, or auxiliary and service personnel, with median incomes of $6000 or less, whereas only 1.3% of professionals (doctors, scientists, administrators) were women. And among academics, women cluster at the lower academic ranks and in the less prestigious institutions. For a more detailed discussion, *see* Ruth Hubbard, "When Women Fill Men's Roles . . . ," *Trends in Biochem. Sci. 1*:N52–53 (1976).

*Figure XXI. Dr. Ellen Swallow Richards conducting water pollution tests with an unidentified assistant (1870's). (M.I.T. Historical Collections)*

# Mary Sue Henifin

# Bibliography: Women, Science, and Health

## Contents

## Introduction

This bibliography is intended to aid students, teachers, and others pursuing research interests regarding women, science, and health. Although far from complete, it is an introduction to the literature and a taking off place for further exploration.

Many of the works referenced are easily obtainable; but a few are included for historical interest, although they are not readily accessible. Finding articles from the 1890's to 1920's on sex and evolution, or feminism and science, for example, makes one realize that interest in these topics is not new.

I have made no attempt to distinguish feminist works from those written from other perspectives: that is best left to the reader. However, I have tried to present the classic feminist works as well as some of the major works from other viewpoints.

Working on a bibliography is both frustrating and exciting. One is never finished—there are always more possibilities to explore, more old works to discover. Suggestions for additional references or format changes for future revisions should be sent to Professor Ruth Hubbard, Biological Laboratories, Harvard University, Cambridge, Mass. 02138.

I would like to thank Ruth Hubbard for her many excellent suggestions and for her encouragement.

## I.   Women and Science

Asterisks (*) indicate books and review articles that I have found particularly useful.

EVOLUTION

Blackwell, Antoinette Brown. "The alleged antagonism between growth and reproduction." *Popular Science Monthly* 29 (Sept. 1874): 606–609.

————. "Letter [on sex and evolution]." *Popular Science Monthly* 31 (July 1876): 362–363.

————. *The Sexes Throughout Nature.* New York: G. P. Putnam's Sons, 1875. Excerpted in *The Feminist Papers.* Edited by Alice Rossi. New York: Bantam Books, 1973; reprinted Westport, Conn.: Hyperion Press, 1978.

Brooks, W. K. "The condition of women from a zoological point of view." *Popular Science Monthly* 34 (June 1879): 145–155.

Bullough, V. L. *The Subordinate Sex.* Urbana, Illinois: University of Illinois Press, 1973.

Campbell, B. *Sexual Selection and the Descent of Man, 1871–1971.* Chicago: Aldine-Atherton, Inc., 1972.

Conway, J. "Stereotypes of femininity in a theory of sexual evolution." *Victorian Studies* 14 (1970): 51.

Cutler, John Henry. *What About Women? An Examination of the Present Characteristics, Nature, Status, and Position of Women as They Have Evolved During this Century.* New York: I. Washburn, 1961.

Darwin, Charles. *The Descent of Man and Selection in Relation to Sex.* New York: D. Appleton & Co., 1872.

————. *The Origin of Species.* 1st Ed. New York: D. Appleton, 1860.

Ehrlich, Carol. "Evolutionism and the place of women in the United States: 1885–1900." In *Women Cross-Culturally.* Edited by Ruby Rohrlich-Leavitt. Chicago: Aldine, 1975.

Ellis, Havelock. *Women and Marriage or Evolution in Sex: Illustrating the Changing Status of Women.* London: William Reeves, 1888.

Engels, Friedrich. *The Origin of the Family, Private Property, and the State.* Edited with an introduction by Eleanor Leacock. New York: International Publishers, 1972.

Gamble, Eliza Burt. *Evolution of Woman: an Inquiry into the Dogma of Her Inferiority to Man.* London, 1849.

Ghiselin, Michael T. *The Economy of Nature and the Evolution of Sex.* Berkeley, California: University of California Press, 1974.

Glaser, Otto Charles. "The constitutional conservatism of women." *Popular Science Monthly* 64 (Sept. 1911): 299–302.

Goldberg, S. *The Inevitability of Patriarchy: Why the Biological Differences Between Men and Women Always Produce Male Domination.* New York: Morrow, 1973.

Hardaker, M. A. "Science and the woman question." *Popular Science Monthly* 35 (March 1882): 577–583.

Holliday, L. *The Violent Sex: Male Psychology and the Evolution of Consciousness*. New York: Bluestocking, 1978.

Hubbard, Ruth. "Have Only Men Evolved?" In *Women Look at Biology Looking at Women*. Edited by Ruth Hubbard, Mary Sue Henifin, and Barbara Fried. Boston, Mass.: G. K. Hall & Co.; Cambridge, Mass., Schenkman Publishing Co., 1979.

Jacoby, Robin Miller. "Science and sex roles in the Victorian era." In *Biology as a Social Weapon*. Edited by the Ann Arbor Science for the People Collective. Minneapolis, Minn: Burgess Publishing Company, 1977.

Johnson, G. W. *The Evolution of Woman from Subjection to Comradeship*. London: Robert Holden, 1926.

Johnson, Robert. *Aggression in Man and Animals*. Philadelphia: W. B. Saunders, 1972.

Leacock, Eleanor. Review of *The Inevitability of Patriarchy*, by Steven Goldberg. *American Anthropologist* 76 (1974): 363–365.

Leakey, Richard E., and Roger Levin. *Origins*. New York: E. P. Dutton, 1977.

Leibowitz, Lila. "Desmond Morris is wrong about breasts, buttocks and body hair." *Psychology Today,* 9 March 1970.

Maccoby, Eleanor E. "Sex in the social order." *Science* 182 (1973): 469–471.

Maudsley, Henry. "Sex in mind and in education." *Popular Science Monthly* 27 (June 1874): 198–214.

Maynard-Smith, J. *Evolution of Sex*. Cambridge, England: Cambridge University Press, 1978.

Morais, Nina. "The women question." *Popular Science Monthly* 21 (May 1882): 70–78.

*Morgan, Elaine. *The Descent of Woman*. New York: Stein and Day, 1972.

*Reed, Evelyn. *Women's Evolution: From Matriarchal Clan To Patriarchal Family*. New York: Pathfinder Press, 1974.

Reyburn, Wallace. *The Inferior Sex*. New York: Pathfinder Press, 1974.

Rhodes, Philip. *Woman: A Biological Study of the Female Role in the Twentieth-Century Society*. London: Corgi, 1969.

*Shields, S. A. "Functionalism, Darwinism, and the psychology of women: A study in social myth." *American Psychologist,* July 1975, pp. 739–754.

Smith, A. Lapthorn. "Higher education of women and race suicide." *Popular Science Monthly* 58 (March 1905): 466–473.

Spencer, Herbert. "On the comparative psychology of the sexes." In his *The Study of Sociology*. New York: D. Appleton, 1875.

——————. *The Principles of Sociology*, Vol. 1, Part III: Westport, Conn: Greenwood Press, 1975. (originally 1897).

Tanner, Nancy, and Adrienne Zihlman. "Women in evolution.

Part I: Innovation and selection in human origins." *Signs* 1 (1976): 585–608.

Tiger, Lionel. *Men in Groups*. New York: Random House, 1969.

Van De Warker, Ely. "The genesis of woman." *Popular Science Monthly* 27 (July 1874): 269–276.

——————. "Sexual cerebration." *Popular Science Monthly* 28 (July 1875): 287–300.

Wallington, Emma. "The physical and intellectual capacities of woman equal to those of man." *Anthropologia* 1 (1874): 552–556.

Washburn, Sherwood L. "Tools and human evolution." *Scientific American* 203 (Sept. 1960): 63–77.

Weitz, Shirley. *Sex Roles: Biological, Psychological and Social Foundations*. Oxford: Oxford University Press, 1977.

Wertheim, W.F. *Evolution and Revolution: The Rising Waves of Emancipation*. London: Penguin Books, 1974.

White, Frances Emily. "Woman's place in nature." *Popular Science Monthly* 28 (Jan. 1875): 292–300.


WOMEN AND MEN: CROSS-CULTURAL PERSPECTIVES


Andreski, Iris, ed. *Old Wives' Tales: Life Stories from Ibibioland*. New York: Schocken Books, 1970.

Ardener, Shirley, ed. *Perceiving Women*. New York: Halstead, 1975.

Bachofen, Johann J. *Myth, Religion and Mother Right*. Translated by Ralph Manheim. Princeton: Princeton University Press, 1967 (originally 1861).

Beidelman, T. O. *The Kaguru, a Matrilineal People of East Africa*. New York: Holt, Rinehart and Winston, 1971.

Bettelheim, Bruno. *Symbolic Wounds: Puberty Rites and the Envious Male*. New York: Collier Books, 1962.

Blurton-Jones, N. G., and M. J. Konner "Sex differences in behavior of London and Bushmen children." In *Comparative Ecology and Behavior of Primates*. Edited by R. P. Michael and J. H. Cook. New York: Academic Press, 1973.

Bricker, Victoria Reifler. "Sex roles in cross-cultural perspective." *American Ethnologist* 2 (Nov. 1975): 4.

Briffault, Robert. *The Mothers*. New York: Grosset and Dunlap, 1963 (originally 1927).

Clignet, Remi. *Many Wives, Many Powers: Authority and Power in Polygynous Families*. Evanston, Illinois: Northwestern University Press, 1970.

Cornelisen, Ann. *Woman of the Shadows*. Boston: Little, Brown, 1976.

Curtin, Katie. *Women in China*. New York: Pathfinder Press, 1975.

Davis, Elizabeth Gould. *The First Sex*. Baltimore: Penguin Books, 1971.

Diner, Helen. *Mothers and Amazons*. New York: The Julia Press, 1965.

Draper, P. "Kung women: Contrasts in sexual egalitarianism in the foraging and sedentary contexts." In *Toward an Anthropology of Women*. Edited by R. Reiter. New York: Monthly Review Press, 1975.

Evans-Pritchard, E. E. *The Position of Women in Primitive Societies and Other Essays in Social Anthropology*. New York: Free Press, 1965.

Farber, Seymour M., and H. L. Roger, eds. *Man and Civilization: The Potential of Women; A Symposium*. New York: Free Press, 1963.

Farmer, Claire R., ed. *Women and Folklore*. Austin, Texas: University of Texas Press, 1975.

Fee, Elizabeth. "The sexual politics of Victorian social anthropology." In *Clio's Consciousness Raised: New Perspectives on the History of Women*. Edited by M. Hartman and L. Banner. New York: Harper and Row, 1974.

Food and Agricultural Organization of the United Nations. *The Missing Half: Woman 1975*. New York: United Nations, 1975.

Ford, Clellan S., and Frank A. Beach. *Patterns of Sexual Behavior*. New York: Harper and Brothers, 1951.

*Friedl, Ernestine. *Women and Men: An Anthropologist's View*. New York: Holt, Rinehart, Winston, 1975.

Gale, Fay, ed. *Women's Role in Aboriginal Society*. Canberra: Australian Institute of Aboriginal Studies, 1970.

Giele, Janet Zoflinger, and Audrey Chapman Smock. *Women: Roles and Status in Eight Countries*. Somerset, New Jersey: Wiley-Interscience, 1977.

Goody, Jack, ed. *The Character of Kinship*. London: Cambridge University Press, 1973.

Hammond, Dorothy, and Alta Jablow. *Women: Their Familial Roles in Traditional Societies*. Menlo Park, California: Cummings Publishing Co., 1975.

――――――. *Women: Their Economic Role in Traditional Societies*. Reading, Mass.: Addison-Wesley Publishing Co., 1973.

Harris, Marvin, *Cannibals and Kings: The Origins of Cultures*. New York: Random House, 1977.

――――――. *Culture, People, Nature*, 2nd edition. New York: Thomas Y. Crowell Co., 1975.

Hogbin, Herbert Ian. *The Island of Menstruating Men*. Scranton, Pa.: Chandler Publishing Co., 1970.

Iglitzin, Lynn B., and Ruth Ross, eds. *Women in the World: A Comparative Study*. Santa Barbara, California: American Bibliographical Center-Clio Press, 1976.

Johnson, Barclay D. "Durkheim on women." In *Woman in a Man-Made*

*World*. 2nd Edition. Edited by Nona Y. Glazer and Helen Waehrer. Chicago: Rand-McNally, 1977.

Joyce, Thomas, ed. *Women of All Nations: A Record of their Characteristics, Habits, Manners, Customs and Influence*. New York: Funk and Wagnalls, 1912.

Kaberry, Phyllis Mary. *Aboriginal Women: Sacred and Profane*. London: Routledge and Sons, 1970.

——. *Women of the Grassfields*. London: H. M. Stationery Office, 1952.

*Kessler, Evelyn S. *Women: An Anthropological View*. New York: Holt, Rinehart and Winston, 1976.

Kessler, S. T., and W. McKenna. *Gender, An Ethnomethodological Approach*. New York: John Wiley and Sons, 1978.

Kolata, Gina B. "Kung hunter-gatherers: Feminism, diet, and birth control." *Science* 185 (1974): 932–934.

*Lamphere, Louise. Review essay on anthropology. *Signs* 2 (Spring 1977): 612–627.

Landes, Ruth. *The City of Women: Negro Women Cult Leaders of Babia, Brazil*. New York: 1947.

——. *The Ojibwa Women*. New York: W. W. Norton and Company, 1971 (originally 1938).

Lapidus, Gail Warshofsky. *Women in Soviet Society: Equality, Development, and Social Change*. Berkeley: University of California Press, 1978.

Leacock, Eleanor Burke. "Class, Commodity and the Status of Women." In *Women Cross-Culturally: Change and Challenge*. Edited by Ruby Rohrlich-Leavitt. Chicago: Aldine, 1975.

——. "Matrilocality in a simple hunting economy." *Southwestern Journal of Anthropology* 2 (1955): 31–47.

——. Review of Louise Spindler, *Menomini Women and Cultural Change*. *American Anthropologist* 65 (1963): 4.

——. "Women, development, and anthropological facts and fictions." *Latin American Perspectives* 4.

——. "Women in egalitarian societies." In *Becoming Visible: Women in European History*. Edited by Bridenthal and Koonz. New York: Houghton Mifflin, 1977.

——, and June Nash. "Ideologies of sex, archetypes and stereotypes." *Annals of the New York Academy of Sciences* 285 (1977).

Lee, Richard B., and Irvin DeVore. *Kalahari Hunter-Gatherers: Studies of the !Kung and their Neighbors*. Cambridge, Mass.: Harvard University Press, 1976.

MacCormack, Carol P. *Biological Events and Cultural Control*. Chicago: University of Chicago Press, 1977.

Malinowski, Bronislaw. *Sex and Depression in Savage Society*. New York: Harcourt, Brace and Co., 1927.

————. *The Sexual Life of Savages in North-Western Melanesia.* New York: Halcyon House, 1941.

*Martin, M. Kay, and Barbara Voorhies. *Female of the Species.* New York: Columbia University Press, 1975.

Mason, Otis T. "Environment in relation to sex in human culture." *Popular Science Monthly* 55 (Feb. 1902): 336–345.

————. *Women's Share in Primitive Culture.* New York: Gordon Press, 1976 (originally 1895).

*Matthiasson, C. J., ed. *Many Sisters: Women in Cross-Cultural Perspective.* New York: Free Press, 1974.

*Mead, Margaret. *Male and Female: A Study of the Sexes in a Changing World.* New York: Dell Publishing Co., 1949.

*————. *Sex and Temperment in Three Primitive Societies.* New York: Dell Publishing Co., 1935.

Minturn, Leigh, and William Lambert. *Mothers of Six Cultures.* New York: John Wiley and Sons, 1964.

Montagu, Ashley. *The Natural Superiority of Women.* New York: Collier Books, 1968.

Murphy, Yolanda, and Robert F. Murphy. *Women of the Forest.* New York: Columbia University Press, 1974.

Paulme, Denise, ed. *Women of Tropical Africa.* Berkeley and Los Angeles: University of California Press, 1963.

Raphael, Dana, ed. *Being Female: Reproduction, Power, and Change.* Chicago: Aldine Publishing Co., 1973.

*Reiter, R., ed. *Toward an Anthropology of Women.* New York: Monthly Review Press, 1975.

Rohrlich-Leavitt, Ruby. *Anthropological Approaches to Women's Status.* New York: Harper and Row, 1975.

*————, ed. *Women Cross-Culturally: Change and Challenge.* Chicago: Aldine, 1975.

Ronhaar, J. H. *Woman in Primitive Motherright Societies.* Holland: J. B. Wolters, 1931.

*Rosaldo, M. Z., and L. Lamphere, eds. *Women, Culture, and Society.* Stanford, California: Stanford University Press, 1974.

Schlegel, Alice. *Male Dominance and Female Autonomy: Domestic Authority in Matrilineal Societies.* New Haven: HRAF Press, 1972.

————, ed. *Sexual Stratification: A Cross-Cultural View.* New York: Columbia University Press, 1977.

Schneider, David M. and Kathleen Gough, eds. *Matrilineal Kinship.* Berkeley and Los Angeles: University of California Press, 1961.

Schreiner, Olive. *Women and Labour.* London: Virago, 1978 (originally 1911).

*Stack, C. B., et al. Review Essay on Anthropology. *Signs* 1 (Autumn 1975): 147–160.

Talbot, D. A. *Women's Mysteries of a Primitive People: The Ibibios of South Africa.* New York: Cassel and Co., 1915.

Thomas, Elizabeth Marshall. *The Gentle People.* New York: Alfred A. Knopf, 1959.

Wolf, Margaret, and Roxanne Witke, eds. *Women in Chinese Society.* Stanford: Stanford University Press, 1975.

SOCIAL BIOLOGY AND PRIMATE STUDIES

Abele, L. G., and S. Gilchrest. "Homosexual rape and sexual selection in ancanthocephalan worms." *Science* 197 (1977): 81–83.

Alper, J., et al. "The implications of sociobiology." *Science* 192 (1976): 424.

Ann Arbor Science for the People Collective: Sociobiology Study Group. "Sociobiology: A New Biological Determinism." In *Biology as a Social Weapon.* Edited by the Ann Arbor Science For the People Collective. Minneapolis, Minn.: Burgess Publishing Co., 1977.

Barash, David B. *Sociobiology and Behavior.* New York: Elsevier North-Holland, 1977.

––––––. "Sociobiology of rape in mallards: Responses of the mated male." *Science* 197 (1977): 788–789.

Breines, Wini, Margaret Cerullo, and Judith Stacey. "Social biology, family studies, and antifeminist backlash." *Feminist Studies* 4 (Feb. 1978): 43–68.

Broad, W. J. "Primate lust." *Science News* 113 (1978): 397.

*Caplan, A. L., ed. *The Sociobiology Debate: Readings on Ethical and Scientific Issues.* New York: Harper and Row, 1978.

Chasin, Barbara. "Sociobiology: A sexist synthesis." *Science for the People* 9 (May-June 1977): 3.

Dawkins, Richard. *The Selfish Gene.* New York: Oxford University Press, 1976.

Doty, Robert L. "A cry for the liberation of the female rodent: Courtship and copulation in rodentia." *Psychological Bulletin* 81 (1974): 159–172.

Goodall, Jane. *In the Shadow of Man.* Boston: Houghton Mifflin, 1971.

Gould, S. J. "Biological potentiality versus biological determinism." *Natural History,* May 1976, p. 12.

––––––. "On human nature" [book review]. *Human Nature* 1 (Oct. 1978): 20–33.

Hamilton, W. D. "Innate social aptitudes of man." In *Biosocial Anthropology.* Edited by R. Fox. London: Halstead Press, 1975.

Hrdy, Sarah Blaffer. *The Langurs of Abu: Female and Male Strategies of Reproduction.* Cambridge, Mass.: Harvard University Press, 1977.

Hubbard, R. "From termite to human behavior" [book review of E. O.

Wilson's *On Human Nature*]. *Psychology Today* 12 (Oct. 1978): 124–134.

Kass-Simon, G. "Female strategies: Adaptations and adaptive animal significance." In *Beyond Intellectual Sexism*. Edited by Joan L. Roberts. New York: David McKay Co., 1976.

Kolata, G. B. "Primate behavior: Sex and the dominant male." *Science* 191 (Jan. 1976): 55–56.

Kurchison, Carl. "Social behavior in infrahuman primates." In *Handbook of Social Psychology*, 2nd ed., Vol. 3. Edited by Carl Kurchison. New York: Russell and Russell, 1968.

Lancaster, Jane B. "In praise of the achieving female monkey." In *The Female Experience*. Edited by Carol Tavris. Del Mar, California: CRM, 1973, pp. 5–9.

*––––––. *Primate Behavior and the Emergence of Human Culture*. New York: Holt, Rinehart, and Winston, 1975.

*Leibowitz, Lila. *Females, Males, Families: A Biosocial Approach*. W. Scituate, Mass.: Duxbury Press, 1978.

Leutenegger, W. "Scaling of sexual dimorphism in body size and breeding system in primates." *Nature* 272 (1978): 610–611.

Lewontin, R. C. "Biological determinism as a social weapon." In *Biology as a Social Weapon*. Edited by the Ann Arbor Science for the People Collective. Minneapolis, Minn: Burgess Publishing Co., 1977.

––––––. "The fallacy of biological determinism." *The Sciences*, March-April 1976, p. 6.

Miller, G. S. "The primate basis of human sexual behavior." *Quarterly Review of Biology* 6 (1931): 379–419.

Mitchell, G., and E. M. Brandt. "Paternal behavior in primates." In *Primate Socialization*. Edited by F. Poirier. New York: Random House, 1972.

Ralls, Katherine. "Mammals in which females are larger than males." *The Quarterly Review of Biology* 51 (June 1976): 245–276.

Reed, Evelyn. *An Answer to the Naked Ape and Other Books on Aggression*. New York: Pathfinder Press, 1971.

––––––. *Sexism and Science*. New York: Pathfinder Press, 1978.

Rossi, Alice S. "A biosocial perspective on parenting." *Daedalus* 106 (Spring 1977): 1–31.

*Sahlins, Marshall. *The Use and Abuse of Biology: An Anthropological Critique of Sociobiology*. Ann Arbor: The University of Michigan Press, 1976.

Sherman, Paul W. "Nepotism and the evolution of alarm calls." *Science* 197 (1977): 1246–1253.

Tavris, Carol. "Male supremacy is on the way out: It was just a phase in the evolution of culture." Interview with Carol Tavris. *Psychology Today*, Jan. 1975.

Tiger, Lionel. "Male dominance? Yes, Alas. A sexist plot? No." *New York Times Magazine,* 25 October 1970.

Warshall, Peter. *"Sociobiology"* [book review]. *CoEvolution Quarterly* 9 (Spring 1976): 88–90.

Wickler, Wolfgang. *The Sexual Code: The Social Behavior of Animals and Men.* Garden City, New York: Doubleday, Anchor Books, 1973.

Williams, George C. *Sex and Evolution.* Princeton: Princeton University Press, 1975.

Williams, Sharlotte Neely. "The limitations of the male/female activity distinction among primates: An extension of Judith K. Brown's 'A note on the division of labor by sex.' " *American Anthropologist* 73 (1971): 805–806.

Wilson, E. O. *On Human Nature.* Cambridge, Mass.: Harvard University Press, 1978.

————. *Sociobiology: The New Synthesis.* Cambridge, Mass.: Belknap Press of Harvard University Press, 1975.

Yerkes, R. M. "Social behavior of chimpanzees: dominance between mates in relation to sexual status." *Journal of Comparative Psychology* 30 (Aug. 1940): 174–186.

————. "Social dominance and sexual status in the Chimpanzee." *Quarterly Review of Biology* 14 (1939): 115–136.

Zihlman, A., and N. Tanner. "Gathering and the hominid adaptation." In *Female Hierarchies.* New York: Harry Frank Guggenheim Foundation Third International Symposium, April 1974.

Zuckerman, Solly. "The social life of primates." *The Realist* 6 (1929): 72–88.

SEX DIFFERENCES:

HORMONAL, GENETIC, PHYSIOLOGICAL, PSYCHOLOGICAL

Arnold, Franz Xaver. *Woman and Man: Their Nature and Mission.* Translated by Rosaleen Brennan. New York: Herder and Herder, 1963.

Bart, Pauline B. "Biological Determinism and Sexism: Is It All in the Ovaries." In *Biology as a Social Weapon.* Edited by the Ann Arbor Science for the People Collective. Minneapolis, Minn: Burgess Publishing Co., 1977.

Belotti, E. G. *What are Little Girls Made Of.* New York: Schocken Books, 1976.

Bernard, Jesse. *Sex Differences: An Overview.* In *Psychology: A Programmed Modular Approach,* Module No. 6. Homewood, Ill.: Learning Systems Co., 1975.

*Bleier, Ruth H. "Brain, Body and Behavior." In *Beyond Intellectual Sexism*. Edited by Joan L. Roberts. New York: David McKay Co., 1976.

————. "Myths of the biological inferiority of women: An exploration of the sociology of biology research." *University of Michigan Papers in Women's Studies* 2 (1976): 39–63.

Burnham, Dorothy. "Biology and gender: False theories about women and blacks." *Freedom Ways* 17 (1977): 8–13.

Burstyn, Joan N. "Brain and intellect: Science, applied to a social issue: 1860-1875." XII Congrés International D'Histoire des Sciences. Paris 1968. *Actes*, Tome IX.

————. "Education and sex: The medical case against higher education for women in England, 1870–1900." *Proceedings American Philosophical Society* 117 (1973): 79.

Chafetz, J. S. *Masculine/Feminine or Human: An Overview of Sex Roles*. Itasca, Ill.: F. E. Peacocks Publishing, 1974.

Chetwynd, J. and O. Hartnett. *The Sex Role System: Psychological and Sociological Perspectives*. London: Routledge, Kegan Paul, 1978.

Clarke, Edward H. *Sex in Education*. Boston: James R. Osgood and Company, 1874.

Delauney, G. "Equality and inequality in sex." *Popular Science Monthly* 20 (Dec. 1881): 184–192.

De Leeuw, Hendrick. *Women: The Dominant Sex*. New York: A. S. Barnes, Thomas Yoseloff, 1957.

Deux, Kay. *The Behavior of Women and Men*. Monterey, California: Brooks/Cole, 1977.

Distant, W. L. "On the mental differences between the sexes." *Journal of the Anthropological Institute* 4 (1875): 78–85.

Duberman, L., ed. *Gender and Sex in Society*. New York: Praeger Publishing, 1975.

Dyer, Ken. "Female athletes are catching up: sex differential in track athletics and swimming performances are declining steadily." *New Scientist*, 22 September 1977, pp. 722-723.

Ellis, Havelock. *Man and Woman: A Study of Human Secondary Sexual Characters*. London: Walter Scott, 1894.

————. "Variation in man and woman." *Popular Science Monthly* (1903).

Fairweather, Hugh. "Sex differences in cognition." *Cognition* 4 (1976): 231–280.

Filene, P. G. *Him Her Self: Sex Roles in Modern America*. New York: New American Library, 1975.

Fried, Barbara. "Boys Will Be Boys Will Be Boys: The Language of Sex and Gender." In *Women Look at Biology Looking at Women*. Edited by Ruth Hubbard, Mary Sue Henifin, and Barbara Fried. Boston,

Mass.: G. K. Hall & Co.; Cambridge, Mass.: Schenkman Publishing Co., 1979.

Friedman, R. C., R. M. Richart, and R. L. Van de Wille, eds. *Sex Differences in Behavior.* A Conference Sponsored by the International Institute for the Study of Human Reproduction. College of Physicians and Surgeons of Columbia University. New York: John Wiley and Sons, 1974.

Frieze, I. H., et al. *Women and Sex Roles: A Social Psychological Perspective.* New York: Norton, 1978.

George, W. L. *Intelligence of Woman.* Boston: Little, Brown, 1916.

Giele, J. Z. *Women and the Future: Changing Sex Roles in Modern America.* New York: The Free Press, 1978.

Gilder, George. *Sexual Suicide.* New York: Quadrangle, 1973.

Gould, S.J. "Women's brains." *Natural History* 87 (1978): 44–50.

Goy, R. W. "Early hormonal influences on the development of sexual and sex-related behavior." In *The Neurosciences: Second Study Program.* Edited by Francis O. Schmitt. New York: Rockefeller University Press, 1970.

*Hubbard, Ruth, and Marian Lowe, eds. *Genes and Gender II: Pitfalls in Research on Sex and Gender.* New York: Gordian Press, 1979.

Hutt, Corinne. *Males and Females: Sex Differentiation—Genetic, Hormonal, Psychological.* Middlesex, England: Penguin, 1972.

Hall, Diana Long. "Biology, sex hormones and sexism in the 1920's." In *Women and Philosophy.* Edited by C. C. Gould and M. W. Wartofsky. New York: G. P. Putnam and Sons, 1976.

————. "Biology, sex hormones and sexism in the 1930's." *Philosophical Forum* 5 (1974): 31–95.

Israel, S. Leon. "The essence of womanhood: Biological inequality." *Obstetrics and Gynecology* 29 (May 1967): 750–756.

*Kaplan, Alexandra G., and Joan P. Bean, eds. *Beyond Sex-Role Stereotypes: Readings Toward a Psychology of Androgyny.* Boston: Little Brown and Co., 1976.

*Lee, Patrick C. and Robert S. Stewart. *Sex Differences: Cultural and Developmental Dimensions.* New York: Urizen Books, 1976.

Maccoby, Eleanor E., ed. *The Development of Sex Differences.* Stanford, California: Stanford University Press, 1966.

————, and Carol Nagy Jacklin. *The Psychology of Sex Differences,* 2 vols. Stanford, California: Stanford University Press, 1974.

McEwen, Bruce S. "Interactions between hormones and nerve tissues." *Scientific American* 235 (July 1976): 48–58.

Money, J., ed. *Sex Research, New Developments.* New York: Holt, Rinehart, and Winston, 1965.

————. and P. Tucher. *Sexual Signatures: On Being a Man or A Woman.* Boston: Little Brown, 1975.

Money, John, and Anke A. Ehrhardt. *Man and Woman, Boy and Girl: Differentiation and Dimorphism of Gender Identity, from Conception to Maturity*. Baltimore: The Johns Hopkins University Press, 1972.

Nemilov, Anton V. *The Biological Tragedy of Woman*. Translated from the Russian by Stephanie Otental. New York: Covici, Friede, 1932.

*Oakley, Ann. *Sex, Gender, and Society*. New York: Harper Colophon Books, 1972.

Pfaff, D., et al. "Neurophysiological analysis of mating behavior response of hormone sensitive reflexes." In *Progress in Physiological Psychology*, Vol. 5. Edited by E. Stellar and J. M. Sprague. New York: Academic Press, 1973.

Ramey, Estelle P. "Sex hormones and executive ability." In *Women and Success*. Edited by Ruth B. Kundsin. New York: William Morrow and Co., 1974, pp. 248–256.

Romanes, George J. "Mental differences in men and women." *Popular Science Monthly* 31 (July 1887): 372–382.

Rosenberg, Miriam. "The biological basis for sex role stereotypes." *Contemporary Psychoanalysis* 9 (1973): 374–391.

Star, S. Leigh. "The Politics of Right and Left: Sex Differences in Hemispheric Asymmetry." In *Women Look at Biology Looking at Women*. Ruth Hubbard, Mary Sue Henifin, and Barbara Fried, eds. Boston, Mass.: G. K. Hall & Co.; Cambridge, Mass.: Schenkman Publishing Co., 1979.

*Tavris, Carol, and Carole Offir. *The Longest War: Sex Differences in Perspective*. New York: Harcourt, Brace and Jovanovich, 1977.

Teitelbaum, M. *Sex Differences: Social and Biological Perspectives*. New York: Anchor Books, 1976.

Thompson, Helen Bradford. *The Mental Traits of Sex*. Chicago: University of Chicago Press, 1903.

Tiger, L. "The possible biological origins of sexual discrimination." *Impact of Science on Society* 20 (1970): 29–44.

*Tobach, Ethel, and Betty Rosoff, eds. *Genes and Gender I: First in a Series on Hereditarianism and Women*. New York: Gordian Press, 1978.

Van Den Berghe, Pierre L. *Another Perspective on Man . . . . and Woman and Child: Age and Sex Differentiation*. Belmont, California: Wadsworth, 1973.

Weitz, Shirley. *Sex Roles: Biological, Psychological and Social Foundations*. New York: Oxford University Press, 1977.

Whalen, Richard E. "Sexual differentiation: Models, methods, and mechanisms." In *Sex Differences in Behavior*. Edited by R. C. Friedman, R. M. Richart, and R. L. Van de Wiele. New York: John Wiley and Sons, 1974.

Whitbeck, Caroline. "Theories of sex differences." in *Women and*

*Philosophy*. Edited by C. C. Gould and M. W. Wartofsky. New York: G. P. Putnam and Sons, 1976.

Young, W. C., R. W. Goy, and C. H. Phoenix. "Hormones and sexual behavior." *Science* 143 (1964): 212–218.


WOMEN IN THE SCIENTIFIC WORKFORCE AND
STRUCTURE OF THE SCIENTIFIC WORKFORCE


Arditti, Rita. "Women in science: Women drink water while men drink wine." *Science For the People* 8 (March 1976): 24.

Auvinen, Rita. "Women and work (II): Social attitudes and women's careers." *Impact of Science on Society* 20 (1970): 85–92.

Bachtold, L. M., and E. E. Werner. "Personality characteristics of women scientists." *Psychological Reports* 31 (1972): 391–396.

Barranger, Elizabeth, et al. "Goals for women in science." *Technology Review*, June 1973, pp. 48–57.

Bayer, A. E. and H. S. Astin. "Sex differences in academic rank and salary among science doctorates in teaching." *Journal of Human Resources* 3 (1968): 191.

––––––, and J. Austic. "Sex differentials in the academic reward system." *Science* 188 (1975): 796.

Bernard, Jessie. *Academic Women*. New York: New American Library, 1974.

Boalt, G., H. Lantz, and H. Herlin. *The Academic Pattern: A Comparison Between Researchers and Non-Researchers, Men and Women*. Stockholm: Almquist and Wiksell, 1973.

Booth, Egbert Perry. *Women at War: Engineering*. London: J. Crowther, 1943.

Bregman, Elsie Oschrin. "A study of vocational traits of women secretaries, lawyers, chemists, statisticians, and women in positions of responsibility in department stores." *Psychological Bulletin* 24 (March 1926).

Chepelinsky, Ana Berta, et al. "Women in chemistry—Part of the 51% minority." *Science for the People* 4 (1972).

Cole, Jonathan. *Woman's Place in the Scientific Community*. New York: John Wiley, 1977.

––––––, and Stephen Cole. *Social Stratification in Science*. Chicago: University of Chicago Press, 1973.

*Conference on the Role of Women in Professional Engineering*. A conference held under the auspices of the Executive office of the president of the U.S. and sponsored by the University of Pittsburgh and the Society of Women Engineers. New York: Society of Women Engineers, 1962.

Couture-Cherki, Monique. "Women in Physics." In *The Radicalization of Science: Ideology of/in the Natural Sciences.* Edited by Hilary Rose and S. Rose. London: The Macmillan Press, 1976. Reprinted in *Ideology of/in the Natural Sciences.* Hilary Rose and S. Rose, eds. Boston, Mass.: G. K. Hall & Co.; Cambridge, Mass.: Schenkman Publishing Co., forthcoming 1979.

Crawford, H. Jean. "Report on the association to aid scientific research by women." *Science* 76 (1932): 492–493.

Dodge, N. T. *Women in the Soviet Economy: Their Role in Economic, Scientific and Technical Development.* Baltimore: Johns Hopkins Press, 1966.

Dumbar, M. "Women in science: how much progress have we really made." *Science Forum* 6 (18 April 1973).

Ernest, J. "Mathematics and sex." *The American Mathematical Monthly* 83 (1976): 595–614.

"Adventures of women in science and labor." *Federation Proceedings* 35 (Sept. 1976): 11.

Ferriman, Annabel. "Women academics publish less than men." *Impact of Science on Society,* UNESCO 25 (1975): 153–154.

Fischer, Ann. "The position of women in anthropology." *American Anthropologist* 70 (1968): 338.

Fox, L. Y. "Women and the career relevance of mathematics and science." *School Science and Mathematics* 76 (1976): 357–365.

Frithiof, P. *Women in Science.* Lund, Sweden: University of Lund, 1967.

Goldman, R. D., R. M. Kaplan, and B. B. Platt. "Sex differences in the relationship of attitudes toward technology to choice of field of study." *Journal of Counseling Psychology* 20 (1973): 412.

Goldsmit, N. F. "Women in science: Symposium and job mart." *Science* 168 (1970): 1124.

Gray, M. "The mathematical education of women." *The American Mathematical Monthly* 84 (1977): 374–377.

————. "Women in mathematics." *The American Mathematical Monthly* 79 (1972): 475–479.

Gruchow, Nancy. "Discrimination: women charge universities, colleges with bias." *Science* 169 (Sept. 25, 1970): 1284–1290.

Haber, Julia Moesel. *Women in the Biological Sciences.* State College, Pennsylvania: Pennsylvania State University Press, 1939.

Halsey, S. D., ed. *Women in Geology.* Canton, N. Y.: Ash Lad Press, 1976.

Handler, Philip. "Women scientists: steps in the right direction." *The Sciences* 18 (March 1978): 6–9.

Hansen, R., and J. Neujahr. "Career development of males and females gifted in science." *The Journal of Educational Research* 68 (1974): 43, 45.

Helson, Ravenna. "Women mathematicians and the creative personality." *Journal of Consulting and Clinical Psychology* 36 (1971): 210–220.

Hopkins, Nancy. "The high price of success in science." *Radcliffe Quarterly* 62 (June 1976): 16–18. See also Replies: *Radcliffe Quarterly* 62 (Sept. 1976): 31–32.

Howell, Mary. "Just like a housewife: Delivering 'human services.'" *Radcliffe Quarterly* 63 (Dec. 1977): 1–4.

Hubbard, Ruth. "Rosalind Franklin and DNA" [book review]. *Signs* 2 (1976): 229–236.

——————. "Sexism in science." *Radcliffe Quarterly* 62 (March 1976): 8–11.

——————. "When women fill men's roles. . . ." *Trends in Biochemical Sciences* 1 (1976): N 52–53.

Hudson, L. "Fertility in the arts and sciences." *Science Studies* 3 (1973): 305.

Inke, Lillian V., and Mildred S. Barker. *Employment Opportunities for Women in Professional Engineering.* Washington, D.C.: Government Printing Office, 1954.

*Jacobs, Karen Folger, and Carol S. Wolman. "Twenty-seven strategies of women academics." In *Women in the Organization.* Edited by Hal H. Frank. Philadelphia: University of Pennsylvania Press, 1977, pp. 256–262.

Kashket, Eva R., et al. "Status of women microbiologists: A study of microbiologists based on objective and subjective criteria is presented." *Science* 183 (1974): 488–494.

Keller, E. F. "Women in science: An analysis of a social problem." *Harvard Magazine* 14 (Oct. 1974).

Kistiakowsky, V. *Women in Engineering, Medicine and Science.* Conference on Women in Science and Engineering. Washington, D.C.: National Research Council, June 1973 (Revised: Sept. 1973).

*Kundsin, Ruth B., ed. "Successful women in the sciences: An analysis of determinants." *Annals of the New York Academy of Science* 208 (1973).

*——————, ed. *Women and Success: The Anatomy of Achievement.* New York: W. Morrow and Co., 1974.

Law, Margaret E., ed. *Goals for Women in Science.* Boston: Women in Science and Engineering, 1972.

Lonsdale, K. "Women in science: reminiscences and reflections." *Impact of Science on Society* 20 (1970): 45.

Lubkin, Gloria. "Women in physics." *Physics Today* 24 (1971): 24.

Maccoby, E. E. "Feminine intellect and the demands of science." *Impact of Science on Society* 20 (1970): 13.

*Malcom, Shirley, Paul Q. Hall, and J. W. Brown. *The Double Bind: The Price of Being a Minority Woman in Science.* AAAS Report No. 76-R-3. Washington, D.C. (April 1976).

Mattfeld, J. A., and C. G. van Aken. *Women and the Scientific Professions: The MIT Symposium on American Women in Science.* Cambridge, Mass.: MIT Press, 1965.

Miller, Helen H. "Science careers for women." *Atlantic Monthly,*
     1957, pp. 123–128.
Motz, A. B. "The roles of the married woman in science." *Marriage and*
     *Family Living* 23 (1961): 374.
Muir, Helen. "Emphasize quality not sex." *Trends in Biochemical Sciences*
     2 (June 1977): N 121–123.
Murphy, Mary Claire, and E. S. Spiro. *Careers for Women in the*
     *Biological Sciences.* U.S. Dept. of Labor, Women's Bureau.
     Washington, D.C.: Government Printing Office, 1961.
Nelson, R. "Women mathematicians and the creative personality."
     *Journal of Consulting and Clinical Psychology* 35 (1970): 210.
Nichols, Roberta. "Women in science and engineering: Are jobs really
     sexless?" *Technology Review* (June 1973).
Parish, J. B., and J. S. Block. "The future for women in science and
     engineering." *Bulletin of the Atomic Scientist* 24 (May 1968): 46.
Perrucci, C. C. "Minority status and the pursuit of professional careers:
     Women in science and engineering." *Social Forces* 49 (1970): 245.
Ramaley, J. *Covert Discrimination and Women in the Sciences.*
     Washington, D.C.: American Association for the Advancement of
     Science Symposium, 1977.
Robinson, Gail. "A woman's place is in the [environmental] movement."
     *Environmental Action* 9 (25 March 1978): 12–23.
Roe, A. "Women in science." *Personnel and Guidance Journal* 44 (1966):
     748.
Rossi, A. S. "Women in science: Why so few?" *Science* 148 (1965):
     1196.
*Ruddick, Sara, and Pamela Daniels, eds. *Working It Out: 23 Women*
     *Writers, Artists, Scientists, and Scholars Talk About Their Lives*
     *and Work.* New York: Pantheon Books, 1977.
Schwartz, B. *Women in Scientific Careers.* Washington, D.C.: National
     Science Foundation, 1961.
Scientific Manpower Commission, AAAS. *Professional Women and*
     *Minorities: A Manpower Data Resource Service.* Washington, D.C.:
     AAAS, 1975.
Shapley, Deborah. "Obstacles to women in science." *Impact of Science*
     *on Society* 25 (1975): 105–114.
Stehelin, Liliane. "Sciences, women and ideology." In *The Radicalization*
     *of Sciences: Ideology of/in the Natural Sciences.* Edited by Hilary
     Rose and S. Rose. London: The Macmillan Press, 1976. Reprinted
     in *Ideology of/in the Natural Sciences.* Hilary Rose and S. Rose,
     eds. Boston, Mass.: G. K. Hall & Co.; Cambridge, Mass.: Schenkman
     Publishing Co., forthcoming 1979.
Tereshkan-Nikolayeva, V. "Women in space." *Impact of Science on*
     *Society* 20 (1970): 5–12.

Tossi, Lucia. "Women's scientific creativity." *Impact of Science on Society* 25 (1975): 105–114.

Vetter, B. M. "The outlook for women in science." *The Science Teacher* 40 (1973): 22.

————. "Women in the natural sciences." *Signs* 1 (Spring 1976): 713–720.

Walberg. "Physics, femininity and creativity." *Developmental Psychology* 1 (1969): 47.

Weisstein, Naomi. "Adventures of a Woman in Science." In *Women Look at Biology Looking at Women*. Edited by Ruth Hubbard, Mary Sue Henifin, and Barbara Fried. Boston, Mass.: G. K. Hall & Co.; Cambridge, Mass.: Schenkman Publishing Co., 1979.

White, Martha S. "Psychological and social barriers to women in science." In *Toward a Sociology of Women*. Edited by C. Safilios-Rothschild. Xerox College Publications, 1972.

White, M. S. "Psychological and social barriers to women in science." *Science* 170 (1970): 413.

Women Authors in Science and Engineering, Boston, Mass. "Goals for women in science." *Technology Review* (June 1973).

Women in Research. "Women in science: working together to vitalize research." *Women's Studies Program: Papers in Women's Studies.* The University of Michigan 1 (1976): 1.

Zinberg, Dorothy. "The past decade for women scientists—win, lose, or draw?" *Trends in Biochemical Sciences* 2 (June 1977): N 123–126.

Zuckerman, Harriet, and Jonathan Cole. "Women in American science." *Minerva* 13 (Spring 1975): 82–102.

## HISTORY AND BIOGRAPHY

*American Men and Women of Science*. New York: R. R. Bowker, 1970.

Basalla, George. "Mary Somerville: A neglected popularizer of science." *New Scientist* (March 1973).

Beecher, Catherine. "Letters to the people on health and happiness." In *Root of Bitterness*. Edited by N. F. Cott. New York: E. P. Dutton & Co., 1972 (originally 1855).

Bennett, A. Hughes. "Hygiene in the higher education of women." *Popular Science Monthly* 26 (Feb. 1880): 519–529.

Bridenthal, R. and Koonz, C. eds. *Becoming Visible: Women in European History*. Boston: Houghton Mifflin Co., 1977.

Brooks, Paul. *The House of Life, Rachael Carson at Work*. Boston: Houghton Mifflin Co., 1972.

Clark, Eugenie. *Lady With a Spear* [autobiography of an ichthyologist]. New York: Harper, 1953.

Clarke, Robert. *Ellen Swallow: The Woman Who Founded Ecology.* Chicago: Follett, 1973.

Colden, Jane. *Jane Colden: Botanic Manuscript.* Limited Edition. New York: Garden Club of Orange and Dutchess Counties, 1963.

Crawford, H. Jean. "Association to aid scientific research by women." *Science* 75 (1932).

Cumings, Elizabeth. "Education as an aid to the health of women." *Popular Science Monthly* 27 (Oct. 1880): 823–827.

Curie, Eve. *Madame Curie, A Biography.* Translated by Vincent Sheean. Garden City, New York: Doubleday, Doran and Co., 1937.

Dakin, Susanna Bryant. *The Perennial Adventure: A Tribute to Alice Eastwood, 1859–1953.* San Francisco: California Academy of Science, 1954.

Destreich-Levie, Nancy. "Women in early American anthropology." In *Pioneers of American Anthropology.* Edited by June Helm. AES Monograph 43. Seattle: University of Washington Press, 1966.

Dubreil-Jacotin, Marie Louise. "Women mathematicians." In *Great Currents of Mathematical Thought.* Edited by F. Le Lionnais. New York: Dover, 1971.

Edwards, Samuel. *The Divine Mistress.* [Emilie du Chatelet: mathematician, geometer, and physicist.] New York: David McKay Co, 1970.

Emberlin, Diane P. *Contributions of Women: Science.* Woodland, California: Dillon Publishers, 1977.

Faucett, Mrs. Henry. *Some Eminent Women of our Times.* London: Macmillan, 1899.

Forster, Emily L. *Analytical Chemistry as a Profession for Women.* London: C. Griffin and Co., 1920.

Gage, Matilda J. *Woman as Inventor.* Fayetteville, N. Y.: New York State Woman Suffrage Association, 1870.

Gamble, E. B. *The Sexes in Science and History.* New York: G. P. Putnam, 1916.

Gardener, Helen Hamilton. *Facts and Fictions of Life.* [Including article on Sex in Brain]. Fenno, 1893.

Golde, Peggy, ed. *Women in the Field: Anthropological Experiences.* Chicago: Aldine Publishing Co., 1970.

Green-Armytage, A. J. *Maids of Honor.* London: W. Blackwood and Sons, 1906.

Hunt, Caroline. *Ellen Richards* [Professor at MIT, founder of home economics]. Boston: Whitcomb and Borrows, 1912.

Ireland, N. O. *Index to Women from Ancient to Modern Times.* Westwood, Mass.: F. W. Faxon Co., 1970.

Kendall, Phebe Mitchell. *Maria Mitchell, Life Letters and Journals.* Boston, 1896.

Lonsdale, Kathleen. "Women in science: Reminiscenses and reflections." *Impact of Science on Society* 20 (1970): 45–59.

Lurie, Alison. "Beatrix Potter: More than just Peter Rabbit." *Ms.,* Sept. 1977, pp. 42–45.

Lurie, N. O. "Women in early American anthropology." In *Pioneers of American Anthropology.* Edited by June Helm. Seattle: University of Washington Press, 1966.

Mead, Margaret. *Blackberry Winter, My Earlier Years.* New York: Morrow, 1972.

————, ed. *Writings of Ruth Benedict: An Anthropologist at Work.* New York: Avon Books, 1959.

Meyer, Gerald Dennis. *Science for Englishwomen 1650–1760: The Telescope, the Microscope, and the Feminine Mind.* Berkeley: University of California Press, 1955.

Michael, Helen Abott. *Studies in Plant and Organic Chemistry and Literary Papers, With Biographical Sketch.* Cambridge, Mass.: Riverside Press, 1907.

Moores, Richard G. "Isabel Bevier, lady with a mission." In *Fields of Rich Toil: The Development of the University of Illinois College of Agriculture.* Urbana: University of Illinois Press, 1970.

*Mosedale, Susan. "Science corrupted. Victorian biologists consider 'the woman question.' " *Journal of the History of Biology* 11 (1978): 1–55.

*Mozans, H. J. [John Augustine Zahn]. *Woman in Science.* Cambridge, Mass: M.I.T. Press, 1974 (originally 1913).

Nearing, Scott, and M. S. Nellie. *Women and Social Progress: A Discussion of the Biologic, Domestic, Industrial and Social Possibilities of American Women.* New York: The Macmillan Co., 1912.

O'Hern, E. "Alice Evans, pioneer microbiologist." *ASM News* 39 (1975): 573.

————. "Cora Mitchell Downs, pioneer microbiologist." *ASM News* 40 1974): 862.

————. "Rebecca Craighill Lancefield, pioneer in microbiology." *ASM News* 41 (1975): 805.

*Osen, Lynn. *Women in Mathematics.* Cambridge, Mass.: MIT Press, 1974.

Owens, Helen B. *Early Scientific Work of Women and Women in Mathematics.* State College, Pennsylvania: Pennsylvania State University Press, 1940.

Popular Science Monthly Editorial. "Biology and women's 'rights.' " *Popular Science Monthly* (Dec., 1878): 201–213.

Reid, Robert. *Marie Curie.* New York: Saturday Review Press, E. P. Dutton and Co., Inc., 1974.

Rizzo, P. V. "Early daughters of Urania: Review of women in astronomy." *Sky and Telescope* (Nov. 1954): 7–9.

Rossiter, M. W. "Women scientists in America before 1920." *American Science* 62 (1974): 312–323.

Sayre, Anne. *Rosalind Franklin and DNA: A Vivid View of What it is*

*Like to Be a Gifted Woman in an Especially Male Profession.*
New York: W. W. Norton & Co., 1975.

*Schacher, Susan, coordinator. *Hypatia's Sisters: Biographies of Women Scientists Past and Present.* Seattle, Washington: Feminists Northwest, 1976.

Singer, Charles. *From Magic to Science.* Chapter 6: The Visions of Hildegard of Bingen, Chapter 7: The Ladies of Salerno. 1928.

————. "The Scientific views and visions of Saint Hildegard, 1098–1180." In *Studies on the History and Method of Science.* Oxford: Clarendon Press, 1917.

Smith-Rosenberg, Carroll, and Charles Rosenberg. "The female animal: Medical and biological views of woman and her role in nineteenth-century America." *The Journal of American History* (Sept. 1973).

Smithsonian Institution. *Women in Science in Nineteenth-Century America* [pamphlet and exhibition catalog]. Washington, D.C.: Smithsonian Institution Press, 1978.

Somerville, Martha. *Personal Recollections from Early Life to Old Age of Mary Somerville.* Boston: Roberts Brothers, 1874.

Sterling, Philip. *Sea and Earth: The Life of Rachel Carson.* New York: Crowell, 1970.

Stern, Madeline B. *We the Women: Career Firsts of Nineteenth-Century America.* New York: Shulte Publishing, 1963.

Strasser, Judy. "Jungle law: Stealing the double helix." *Science for the People* 8 (Sept./Oct. 1976): 29–31.

Trescott, Martha B. "Julia B. Hall and aluminum." *Journal of Chemical Education* 54 (1977): 24.

Truman, Margaret. *Women of Courage.* New York: Morrow, 1976.

Vicinus, Martha. *Suffer and Be Still: Women in the Victorian Age.* Bloomington, Indiana: University of Indiana Press, 1972.

Visher, Stephen S. "Women starred in 'American Men of Science, 1903–1943.'" In *Scientists Starred.* Edited by S. Visher Stephen. Baltimore: Johns Hopkins Press, 1947.

Wade, Ira O. *Voltaire and Madame du Chatelet: An Essay on the Intellectual Activity at Circy.* Princeton, New Jersey: Princeton University Press, 1941.

Wade, Nicholas. "Discovery of pulsars: a graduate student's story." *Science* 189 (1975): 358–364.

Willard, Mary Louisa. *Pioneer Women in Chemistry.* State College, Pennsylvania: Pennsylvania State University Press, 1940.

William, Henry Smith. *The Great Astronomers.* New York: Simon and Schuster, 1930.

Williams, Margot, and Paul Elliot. "Maria Martin: The brush behind Audubon's birds." *Ms.* 5 (1977): 14–18.

Wilson, Carol Green. *Alice Eastwood's Wonderland, The Adventures of a Botanist.* San Francisco, California: California Academy of Science, 1955.

Wright, Helen. *Sweeper of the Sky: The Life of Maria Mitchell, First Astronomer in America.* New York: Macmillan Co., 1949.

Wupperman, Alice. "Women in 'American Men of Science.' " *Journal of Chemical Education* (1941): 120–121.

Yost, Edna. *American Women of Science.* New York: Fred A. Stokes, 1943.

————. *Women of Modern Science.* New York: Dodd, Mead, and Co., 1959.


FUTURES


Birkby, Phyllis, and Leslie Weisman. "Patritecture and feminist fantasies." *Liberation Magazine* (1976): 48–52.

Bryant, Dorothy. *The Kin of Ata are Waiting for You.* San Francisco: Moon Books, Random House, 1976.

LeGuin, Ursula. *The Dispossessed.* New York: Avon Books, 1974.

————. *The Left Hand of Darkness.* New York: Ace Books, 1969.

————. "The space crone." *The CoEvolution Quarterly* (Summer 1976): 108–110.

Madsen, Catherine. "Commodore Bork and the compost." *Women: A Journal of Liberation* 5 (1977): 32–35.

Piercy, Marge. *Woman on the Edge of Time.* New York: Alfred A. Knopf, 1976.


GENERAL


*The Ann Arbor Science for the People Collective. *Biology as a Social Weapon.* Minneapolis, Minn: Burgess Publishing Co., 1977.

Arditti, Rita. "Women's biology in a man's world: some issues and questions." *Science for the People* 4 (July 1973).

————, Pat Brennan, and Steve Cavrak. *Science and Liberation.* Boston: South End Press, 1979.

Beatty, Jerome. *The Girls We Leave Behind: A Terribly Scientific Study of American Women at Home.* Garden City, New York: Doubleday, 1963.

Berger, Peter L., and Thomas Luckmann. *The Social Construction of Reality: A Treatise in the Sociology of Knowledge.* Garden City, New York: Anchor Books, Doubleday and Company, 1966.

Bird, Caroline. *Born Female.* New York: David McKay Co., 1968.

Bleier, Ruth. "Difficulties of detecting sexist biases in the biological sciences." *Signs* 4 (1978): 159–162.

Braxton, Bernard. *Women, Sex and Race*. Washington, D.C.: Verta Press, 1973.

Costa, M. Dalla, and S. James. *The Power of Women and the Subversion of the Community*. Bristol: Falling Wall Press, 1972.

*Cott, Nancy F., ed. *Root of Bitterness: Documents of the Social History of American Women*. New York: E. P. Dutton Co., 1972.

*Daly, M. *Beyond God the Father: Toward a Philosophy of Women's Liberation*. Boston: Beacon Press, 1973.

——————. *Gyn/Ecology: The Metaphysics of Radical Feminism*. Boston: Beacon Press, 1978.

*de Beauvoir, Simone. *The Second Sex*. New York: Alfred A. Knopf, 1953; Bantam Books, 1961.

Figes, Eva. *Patriarchal Attitudes*. New York: Fawcett, Premier Books, 1970.

*Firestone, Shulamith. *The Dialectic of Sex*. New York: Bantam Books, 1970.

Freeman, J., ed. *Women: A Feminist Perspective*. Palo Alto, California: Mayfield Publishing Co., 1975.

*Gornick, Vivian and B. K. Moran. *Woman in Sexist Society*. New York: Basic Books, 1971.

*Griffin, Susan. *Woman and Nature: The Roaring Inside Her*. New York: Harper and Row, 1978.

Hall, Diana Long. "The social implications of the scientific study of sex." *The Scholar and the Feminist IV: Connecting Theory, Practice and Values*. A Conference sponsored by the Barnard College Women's Center, Papers from the Morning Session, April 23, 1977. The Women's Center, Barnard College.

*Herschberger, Ruth. *Adam's Rib*. New York: Har/Row Books, 1970 (originally 1948).

Hochschild, Arlie Russel. "A review of sex role research." *American Journal of Sociology* 78 (Jan. 1973): 1011–1029.

*Hubbard, Ruth, Mary Sue Henifin, and Barbara Fried, editors. *Women Look at Biology Looking at Women: A Collection of Feminist Critiques*. Boston, Mass.: G. K. Hall & Co.; Cambridge, Mass.: Schenkman Publishing Co., 1979.

Lipman-Blumen, J. "How ideology shapes women's lives." *Scientific American* 226 (1972): 34–42.

Lowie, Robert H., and Leta Stetter Hollingwort. "Science and feminism." *Scientific Monthly* (Sept. 1916).

Mill, John Stuart. *On the Subjection of Women*. New York: Collectors Editions, Source Book Press, 1970 (originally 1869).

——————, and Harriet Taylor Mill. *Essays on Sex Equality*. Compiled by Alice S. Rossi. Chicago: University of Chicago Press, 1970.

Millman, Marcia, and Rosabeth Moss Kanter. *Another Voice*. Garden City, N. Y.: Anchor Books, 1975.

Morgan, Robin, ed. *Sisterhood Is Powerful*. New York: Vintage, 1970.

Ramey, Estelle. "A feminist talks to men." *Johns Hopkins Magazine* (Sept. 1973).

————. "An Interview with Dr. Estelle Ramey." *Perspectives in Biology and Medicine* 14 (1971).

Redstockings. *Feminist Revolution.* New York: Random House, 1978.

*Roberts, Joan L. *Beyond Intellectual Sexism: A New Woman, A New Reality.* New York: David McKay Co., 1976.

*Rose, Hilary, and S. Rose, eds. *The Political Economy of Science.* London: The Macmillan Press, 1976.

*————, eds. *The Radicalization of Science: Ideology of/in the Natural Sciences.* London: The Macmillan Press, 1976.

*————, eds. *Ideology of/in the Natural Sciences.* Boston, Mass.: G. K. Hall & Co.; Cambridge, Mass.: Schenkman Publishing Co., forthcoming 1979.

*Rossi, Alice, ed. *The Feminist Papers.* New York: Bantam Books, 1973.

Roszak, Betty, and Theodore Roszak, (eds.) *Masculine/Feminine.* New York: Harper, 1969.

Rousseau, J.-J. *Emile: or a Treatise on Education.* Translated by W. H. Payne, New York: Appleton, 1899.

Ryan, Mary P. *Womanhood in America: From Colonial Times to the Present.* New York: Franklin Watts, 1975.

Seavey, Carol A., Phyllis Katz, A. Zalk, and Sue Rosenberg. "Baby X: The effect of gender labels on adult responses to infants." *Sex Roles* 1 (June 1975): 103–109.

*Signs* 4:1 ["Women, Science, and Society"] (1978): 1–216.

Stoll, C. S. *Sexism: Scientific Debates.* Menlo Park, California: Addison-Wesley, 1973.

Tobias, Sheila. *Overcoming Math Anxiety.* New York: W. W. Norton, 1978.

*Wollstonecraft, Mary. *A Vindication of the Rights of Women.* Dublin: James Moore, 1793.

*Woolf, Virginia. "Professions for Women." In *The Death of the Moth and Other Essays.* New York: Harcourt Brace, 1942; Harvest, 1974.

*————. *A Room of One's Own.* New York: *Penguin,* 1964 (originally 1928).

*————. *Three Guineas.* London: Hogarth Press, 1968 (originally 1938).

Wortig, Rochelle. "The acceptance of the concept of the maternal role by behavioral scientists: Its effect on women." In *The Women's Movement.* Edited by Helen Wortig and Clara Rabinowitz. New York: AMS Press, 1972.

## II.   Women and Health

WOMEN AND THE MEDICAL PROFESSION

American Psychiatric Association. *Women in Psychiatry*. Washington, D.C.: APA, 1973.

Ashley, JoAnn. *Hospitals, Paternalism and the Role of the Nurse*. New York: Teachers College Press, 1976.

Beshiri, Patricia H. *The Woman Doctor: Her Career in Modern Medicine*. New York: Cowles Book Co., 1969.

Brown, Carol A. "Women workers in the health service industry." *International Journal of Health Services* 5 (1975): 173–184.

Bullough, Bonnie. "Barriers to the nurse practitioner movement." *International Journal of Health Services* 5 (1975): 225–234.

Campbell, J. E. "Women dentists: An untapped resource." *Journal of the American College of Dentistry* 36 (1970): 265–269.

*Campbell, Margaret [Mary C. Howell]. *"Why Would a Girl Go Into Medicine?" Medical Education in the United States: A Guide for Women*. Old Westbury, N. Y.: The Feminist Press, 1973.

Cannings, Kathleen, and William Lazonick. "The Development of the nursing labor force in the United States: A basic analysis." *International Journal of Health Services* 5 (1975): 185–216.

Chaff, S. L., et al. *Women in Medicine: A Bibliography of the Literature on Women Physicians*. Metuchen, N. J.: Scarecrow Press, 1977.

Conference on Meeting Medical Manpower Needs. *The Fuller Utilization of the Woman Physician: Report*. Washington, D.C.: American Medical Women's Association, 1968.

*Corea, Gena. *The Hidden Malpractice: How American Medicine Treats Women as Patients and Professionals*. New York: Wm. Morrow and Co., 1977.

Coste, Chris. "Women in medicine: Progress and prejudice." *The New Physician* 24 (1975): 25–31.

Fee, Elizabeth. "Women and health care: A comparison of theories." *International Journal of Health Services* 5 (1975): 397–415.

Feldman, Jacqueline. "The savant and the midwife." *Impact of Science on Society* 24 (1975): 105–115.

Fitzpatrick, M. Louise. "Nursing." *Signs* 2 (Summer, 1977): 818–834.

Grissum, M., and C. Spengler. *Womanpower and Health Care*. Boston: Little Brown and Co., 1976.

Hare, Daphne. "The victim is guilty." *Federation Proceedings* 35 (1975): 2223–2225.

Haseltine, Florence, and Yvonne Yaw. *Woman Doctor: The Internship of a Modern Woman*. Boston, Mass: Houghton Mifflin Co., 1976.

Jacobson, Arthur C. "A medical view of women's lib." *Medical Times* 180 (July 1972).

Jefferys, Margot. *Women in Medicine: The Results of an Inquiry Conducted by the Medical Practitioners' Union in 1962–63.* London: Office of Health Economics, 1966.

Linn, Edwin. "Women dentists: Career and family." *Social Problems* 18 (1971): 394–395.

————. "Women dentists: Some circumstances about their choice of a career." *Journal of the Canadian Dental Association* 10 (1972): 364–409.

Lopate, C. *Women in Medicine.* Baltimore: Johns Hopkins Press, 1964.

*Lorber, Judith. "Women and medical sociology: Invisible professionals and ubiquitous patients. In *Another Voice.* Edited by M. Millmann and R. M. Kanter. Garden City, N. Y.: Anchor Books, 1975, pp. 75–105.

Lutzker, Edythe. *Women Gain a Place in Medicine.* New York: McGraw Hill, 1969.

Marieskind, Helen. "The women's health movement." *International Journal of Health Services* 5 (1975): 217–224.

Melnick, Vijaya, and Franklin D. Hamilton. *Minorities in Science: The Challenge for Change in Biomedicine.* New York: Plenum Press, 1977.

Nathanson, C. A. "Illness and the feminine role: A theoretical review." *Social Science Medicine* 9 (1975): 57–62.

Romney, Seymour L., et al. *Gynecology and Obstetrics: The Health Care of Women.* New York: McGraw-Hill, 1975.

Scully, Diane, and Pauline Bart. "A funny thing happened on the way to the orifice: Women in gynecology textbooks." *American Journal of Sociology* 78 (1973): 1045.

Slade, Margot. "The women in white." *The New Physician* 24 (1975): 34–35.

Spieler, Carolyn, ed. *Women in Medicine—1976.* New York: Independent Publishers Group, 1977.

U.S. Dept. of Health, Education, and Welfare. *Minorities and Women in the Health Fields: Applicants, Students and Workers* (DHEW Pub. #MRA 76-22). Washington, D.C.: Government Printing Office, 1975.

U.S. Dept. of Health, Education, and Welfare. Women's Action Program, Office of Special Concerns. *An Exploratory Study of Women in the Health Professions Schools, Vol. IV, Dentistry.* California Urban and Rural Systems Association, 1976.

*Walsh, Mary Roth. *Doctors Wanted: No Women Need Apply. Sexual Barriers in the Medical Profession, 1835–1975.* New Haven: Yale University Press, 1977.

————. "The Quirks of a Woman's Brain." In *Women Look at Biology*

*Looking at Women.* Ruth Hubbard, Mary Sue Henifin, and Barbara
Fried, eds. Boston, Mass.: G. K. Hall & Co.; Cambridge, Mass.:
Schenkman Publishing Co., 1979.
Williams, J. J. "The woman physician's dilemma." *Journal of Sociological
Issues* 6 (1950): 38.

MENSTRUATION AND MENOPAUSE

Chadwick, M. "The psychological effects of menstruation." *Nervous and
Mental Disease Monographs,* Series 56 (1952).
Clay, Vidal S. *Women: Menopause and Middle Age.* Pittsburgh, Penn-
sylvania: Know Inc., 1977.
Cooper, W. *Don't Change: A Biological Revolution for Women*
[menopause]. New York: Stein and Day, 1975.
Coppen, A., and N. Kessal. "Menstruation and personality." *British Journal
of Psychiatry* 109 (1963): 711–721.
Culpepper, Emily. "Exploring Menstrual Attitudes." In *Women Look at
Biology Looking at Women.* Ruth Hubbard, Mary Sue Henifin, and
Barbara Fried, eds. Boston, Mass.: G. K. Hall & Co.; Cambridge,
Mass.: Schenkman Publishing Co., 1979.
————. *Period Piece* [10 minute, 16 mm color film]. Available
through Insight Exchange. P.O. Box 42594, San Francisco,
California 94101.
*Dalton, Katharina. *The Menstrual Cycle.* New York: Warner Paperback
Library, 1972.
*Delany, Janice, Mary Jane Lupton, and Emily Toth. *The Curse: A Cultural
History of Menstruation.* New York: A Mentor Book, New American
Library, 1976.
Duke, Alexander. "The use of the sponge pessary during menstruation."
*The Medical Press* 58 (Nov. 1894).
Edgar, J. Clifton. "Bathing during the menstrual period." *American
Journal of Obstetrics* 1 (1904).
Ernster, Virginia L. "American menstrual expressions." *Sex Roles* 1
(March 1975): 3–13.
Flint, Marsha. "The menopause—reward or punishment?" *Psychosomatics*
16 (1975): 4.
Frisch, Rose E. "Demographic implications of the biological determinants
of female fecundity." *Social Biology* 22 (1975): 17–22.
————, and Janet W. McArthur. "Menstrual cycles: fatness as a
determinant of minimum weight for height necessary for their
maintenance and onset. *Science* 185 (1974): 949–951.
————, and R. Revelle. "Height and weight at menarche and a
hypothesis of menarche." *Archives of Diseases of Childhood* 46
(1971): 695–701.

Golub, Sharon. "The magnitude of premenstrual anxiety and depression." *Psychosomatic Medicine* 38 (1976): 1.

Grossman, Marlyn, and Pauline Bart. "Taking the Men Out of Menopause." In *Women Look at Biology Looking at Women*. Ruth Hubbard, Mary Sue Henifin, and Barbara Fried, eds. Boston, Mass.: G. K. Hall &. Co.; Cambridge, Mass.: Schenkman Publishing Co., 1979.

Gruba, Glen H., and Michael Rohrbaugn. "MMPI correlates of menstrual distress." *Psychosomatic Medicine* (1975).

Janowsky, D., R. Gorney, and B. Kelly. "The curse: Vicissitudes and variations of the female fertility cycle. Part I: Psychiatric aspects. *Psychosomatics* 7 (1966): 242–246.

Levitt, E. E., and B. Lubin. "Some personality factors associated with menstrual complaints and menstrual attitude." *Journal of Psychosomatic Research* 11 (1967): 267–270.

Mandell, Arnold J., and Mary Mandell. "Suicide and the menstrual cycle." *Journal of the American Medical Association* 200 (1967): 792–793.

Merriman, Georgia. "Do women require mental and bodily rest during menstruation." *Columbus Medical Journal* 13 (1894).

Nachtigall, L., with T. Heilman. *The Nachtigall Report* [menopause and estrogen therapy]. New York: G. P. Putman and Sons, 1977.

Newman, J. H. *Our Own Harms: The Startling Facts Behind Menopause and Estrogen Therapy*. Newport Beach, California: Quail Street Publishing Co., 1974.

Reitz, Rosetta. *Menopause: A Positive Approach*. Radnor, Pennsylvania: Chilton Book Co., 1977.

Rose, Louise. *The Menopause Book*. New York: Hawthorn, 1977.

*Seaman, Barbara, and Gideon Seaman. *Women and the Crisis in Sex Hormones*. New York: Rawson Associates, 1977.

Stern, K., and M. Prado. "Personality studies in menopausal women." *American Journal of Psychiatry* 103 (1946): 358–368.

Svennerva, Sven. "Dysmenorrhoea and Absenteeism: Some Gynaecologic and Medico-Social Aspects." Translated by L. James Brown. *Acta Obstetrica et Gynecologica Scandinavica*, Vol. 38. Lund, 1950.

*Weideger, Paula. *Menstruation and Menopause: The Physiology and Psychology, The Myth and the Reality*. New York: Alfred A. Knopf, 1976.

Wheat, Valerie. "The red rains: A period piece." *Crysalis* 1 (Spring 1977).

*Womanspirit Magazine* 1 (WinterSolstice, 1974): 2 [Box 263 Wolf Creek, Oregon, 97497].

BIRTH CONTROL, ABORTION, AND STERILIZATION

Arnstein, Helene S. *What Every Woman Needs to Know About Abortion*. New York: Charles Scribner, 1973.

Banks, Joseph Ambrose, and Banks, Olive. *Feminism and Family Planning in Victorian England*. New York: Schocken Books, 1964.

Barr, S. J. *A Woman's Choice* [abortion]. New York: Rawson Associates Publishing, 1977.

Beral, Valerie. "Cardiovascular-disease mortality trends and oral-contraceptive use in young women." *The Lancet,* 13 Nov. 1976, pp. 1047–1051.

Birdsall, Nancy. "Women and population studies." *Signs* 1 (Spring 1976): 699–712.

Carter, Luther J. "New feminism: potent force in birth control policy." *Science* 167 (1970): 1234–1236.

Cohen, M., T. Nagel, and T. Scanlon. *The Rights and Wrongs of Abortion.* Princeton, N. J.: Princeton University Press, 1974.

Cone, Jim. "Forced sterilization and the poor." *Synapse* (Spring 1975).

Devereux, George. *A Study of Abortion in Primitive Societies.* New York: Julian Press, 1955.

Draper, Elizabeth. *Birth Control in the Modern World.* Baltimore: Penguin Books, 1965.

Dreifus, Claudia. "Sterilizing the poor." *The Progressive* 39 (Dec. 1975): 12.

Dumund, Don E. "The limitation of human population: A natural history." *Science* 187 (1974): 713–721.

Eliot, Johan N. "Fertility control and coercion." *Family Planning Perspectives* 5 (Summer 1973): 3.

Francke, L. B. *The Ambivalence of Abortion.* New York: Random House, 1978.

Frisch, Rose E. "Population, food intake, and fertility." *Science* 199 (Jan. 1978): 22–33.

Garfink, C., and H. Pizer. *The New Birth Control Program: The Safe and Sure Method of Natural Contraception.* New York: A Bolder Book, 1977.

Gebhard, Paul H. *Pregnancy, Birth, and Abortion.* New York: John Wiley and Sons, Inc., 1958.

Gillette, Paul. *Vasectomy.* New York: Paperback Library, 1972.

Glass, Robert H., and Nathan G. Kase. *Women's Choice: A Guide to Contraception, Fertility, Abortion, and Menopause.* New York: Basic Books, 1970.

Gordon, Linda. "The politics of birth control, 1920–1940: The impact of professionals." *International Journal of Health Services* 5 (1975): 253–278.

*————. *Woman's Body Woman's Right: A Social History of Birth Control in America.* New York: Grossman Publishers, 1976.

Guren, D., and N. Gillette. *The Ovulation Method: Cycles of Fertility—A Natural Birth Control Method that Outdates Rhythm.* Ovulation

Method Teachers Association, 760 Aldrich Road, Bellingham, Washington 98225, 1978.

*Health/Pac Bulletin #62.* "Sterilization of women: The facts." (Jan/Feb. 1975).

Hennekens, Charles, and Brian MacMahon. "Oral contraceptives and myocardial infarction." *New England Journal of Medicine* 296 (May 1977): 1166–1167.

Himes, Norman E. *A Medical History of Contraception.* New York: Schocken Books, 1970.

––––––. "Note on the early history of contraception in America." *New England Journal of Medicine* 205 (1931).

Kaye, Archiprete, et al. *The Abortion Business: A Report on Free-Standing Abortion Clinics, 1975.* Women's Research Action Project, Box 119, Porter Square Station, Cambridge, Mass., 02140.

Kennedy, David M. *Birth Control in America: The Career of Margaret Sanger.* New Haven: Yale University Press, 1970.

Knight, Patricia. "Women and abortion in Victorian and Edwardian England." *History Workshop* 4 (Autumn, 1977): 57–69.

Knodel, John. "Breast feeding and population growth." *Science* 198 (1977): 1111–1115.

Lacey, Louise. *Lunaception.* New York: Coward, McCann and Geoghegan, 1975.

Lader, L. *Abortion.* Boston: Beacon Press, 1966.

Loebel, S. *Conception, Contraception: A New Look.* New York: McGraw Hill, 1974.

Luker, Kristin. *Taking Chances: Abortion and the Decision Not to Contracept.* Berkeley: University of California Press, 1975.

Mass, Bonnie. *Population Target: The Political Economy of Population Control in Latin America.* Ontario, Canada: Charters Publishing Co., 1976.

May, Robert E. "Human reproduction reconsidered: Natural regulation of fertility." *Nature* 272 (April 1978): 491–495.

Mohr, J. C. *Abortion in America: The Origins and Evolution of National Policy.* New York: Oxford University Press, 1978.

Newland, Kathleen. *Women and Population Growth: Choice Beyond Childbearing.* Washington, D.C.: Worldwatch Institute, 1977.

Noiziger, M. *A Cooperative Method of Birth Control,* 2nd ed., rev. Summertown, Tenn.: The Book Publishing Co., 1976.

Notman, M. T., and C. C. Nadelson. *The Woman Patient: Medical and Psychological Interfaces.* Vol. I: *Sexual and Reproductive Aspects of Women's Health Care.* New York: Plenum Press, 1978.

Peel, J., M. Potts, and P. Diggory. *Abortion.* Cambridge, England: Cambridge University Press, 1977.

Perkins, R. L. *Abortion: Pro and Con.* Cambridge, Mass.: Schenkman, 1974.

Piers, M. W. *Infanticide, Past and Present*. New York: W. W. Norton and Co., 1978.

*Reed, James. *From Private Vice to Public Virtue: The Birth Control Movement and American Society Since 1830*. New York: Basic Books, 1978.

Report Prepared for the American Friends Service Committee. *Who Shall Live? Man's Control Over Birth and Death*. New York: Hill and Wang, 1970.

Ris, H. W. "The essential emancipation: The control of reproduction." In *Beyond Intellectual Sexism*. Edited by Joan L. Roberts. New York: David McKay Co., 1976.

Rosen, Harold. *Abortion in America*. Boston: Beacon Press, 1967.

Royal College of General Practitioners' Oral Contraceptive Users Study. "Mortality among oral-contraceptive users." *The Lancet*, 8 Oct. 1977, pp. 728–733.

Schulder, Diane, and Florynce Kennedy. *Abortion Rap: Testimony by Women Who Have Suffered the Consequences of Restrictive Abortion Laws*. New York: McGraw Hill Book Company, 1971.

*Seaman, Barbara. *The Doctors' Case Against the Pill*. New York: Peter H. Wyden, 1969.

Shapiro, Howard I. *The Birth Control Book*. New York: St. Martin's Press, 1977.

Short, R. V. "The evolution of human reproduction." *Proceedings of the Royal Society, London* Series B 195 (December 1976): 3–24.

Skowronski, M. *Abortion and Alternatives*. Millbrae, California: Les Femmes Publishing, 1977.

Tucker, Tarvez. *Birth Control*. New Canaan, Conn.: Tobey Publishing Co., 1975.

Weissman, Steve. "Why the population bomb is a Rockefeller baby." *Ramparts* 8 (May, 1970): 42–47.

Wharton, Lawrence Richardson. *The Ovarian Hormones: Safety of the Pill, Babies after Fifty*. Springfield, Illinois: C. C. Thomas, 1967.

Wylie, Evan McLeod. *All About Voluntary Sterilization*. New York: Berkley Medallion Books, 1977.

## CHILDBIRTH AND MOTHERHOOD

*Arms, Suzanne. *Immaculate Deception: A New Look at Women and Childbirth in America*. San Francisco: San Francisco Book Company/Houghton Mifflin, 1975.

Ashdown-Sharp, P. *A Guide to Pregnancy and Parenthood for Women on Their Own*. New York: Vintage Books, 1977.

Aveling, James H. *English Midwives: Their History and Prospects*. London: Hugh K. Elliott, 1967.

Bean, C. A. *Labor and Delivery.* Garden City, N.Y.: Doubleday and Co., 1977.

——————. *Methods of Childbirth.* Garden City, N.Y.: Dolphin Books, 1974.

*Bernard, Jessie. *The Future of Motherhood.* New York: Penguin Books, 1974.

Bibring, G. L. "Some considerations of the psychological processes in pregnancy." In *The Psychoanalytic Study of the Child.* New York: International Press, 1959.

Bing, Elizabeth. *The Adventure of Birth.* New York: Ace Books, 1970.

——————. *Six Practical Lessons for an Easier Childbirth: The Lamaze Method.* New York: Bantam Books, 1967, 1977.

——————, and Libby Colman. *Making Love During Pregnancy.* New York: Bantam Books, 1977.

*Boston Women's Health Book Collective, Inc. *Ourselves and our Children: A Book by and for Parents.* New York: Random House, 1978.

Brack, Datha C. "Displaced—the midwife by the male physician." *Women and Health* 1 (Nov.-Dec., 1976): 18–24.

Brack, Datha Clapper. "Displaced—the Midwife by the Male Physician." In *Women Look at Biology Looking at Women.* R. Hubbard, M. S. Henifin and B. Fried, eds. Boston, Mass.: G. K. Hall & Co.; Cambridge, Mass.: Schenkman Publishing Co., 1979.

——————. "Social forces, feminism and breastfeeding." *Nursing Outlook* 23 (Sept., 1975): 9.

Brennan, Barbara, and Joan Rattner Heilman. *The Complete Book of Midwifery.* New York: E. P. Dutton and Co., 1976.

Brockbank, William. "Mrs. Jane Sharp's advice to midwives." *Medical History* 2 (1958).

Brook, Danae. *Naturebirth.* New York: Pantheon Books, 1976.

Chabon, I. *Awake and Aware* [drugless childbirth]. New York: Dell Publishing Co., 1966.

Clyne, Douglas G. *A Concise Textbook for Midwives.* London: Faber and Faber, 1975.

Colman, Arthur and Libby. *Pregnancy: The Psychological Experience.* New York: Bantam Books, 1977.

Cutter, Irving S., and Henry R. Viets. *A Short History of Midwifery.* Philadelphia: Saunders, 1964.

Daffenbarger, R. S. "The picture puzzle of the postpartum psychosis." *Journal of Chronic Diseases* 13 (1961): 171–173.

Demeter, Anna. *Legal Kidnapping. A Mother's Account of What Happens to a Family When the Father Kidnaps Two Children.* Boston: Beacon Press, 1977.

Devitt, Neal. "The transition from home to hospital birth in the United States, 1930–1960." *Birth and the Family Journal* 1 (Summer 1977): 47–58.

Dick-Read, Grantly. *Childbirth Without Fear: The Principles and Practices of Natural Childbirth*. New York: Dell, 1962 (originally 1944).

Dilfer, Carol Stahman. *Your Baby, Your Body: Fitness During Pregnancy*. New York: Crown Publishing, 1977.

Donegan, Jane B. *Women and Men Midwives: Medicine, Morality and Misogyny in Early America*. Westport, Conn.: Greenwood Press, 1978.

Donovan, B. *The Caesarean Experience*. Boston: Beacon Press, 1977.

Dunn, Peter M. "Obstetric delivery today: For better or for worse." *The Lancet,* 10 April 1976.

Eiger, Marvin S., and Sally Wendkos Olds. *The Complete Book of Breastfeeding*. New York: Bantam Books, 1972.

Elkins, Valmai Howe. *The Rights of the Pregnant Parent*. New York: Two Continents Publishing Co., 1976.

Emmons, Arthur Brewster, and James Lincoln Huntington. "A review of the midwife situation." *Boston Medical and Surgical Journal* 164 (1911).

Ewy, Donna, and Rodger Ewy. *Preparation for Childbirth: A Lamaze Guide*. Garden City, N.Y.: Doubleday Books, 1975.

Feldman, S. *Choices in Childbirth*. New York: Grosset and Dunlap, 1978.

Forbes, Thomas Rogers. *The Midwife and the Witch*. New Haven: Yale University Press, 1966.

——————. "Midwifery and witchcraft." *Journal of the History of Medicine* 17 (1962).

——————. "The regulation of English midwives in the sixteenth and seventeenth centuries." *Medical History* 8 (1964).

Fraiberg, Selma. *Every Child's Birthright: In Defense of Motherhood*. New York: Basic Books, 1977.

Francoeur, Robert. *Utopian Motherhood—New Trends in Human Reproduction*. Garden City, N.Y.: Doubleday and Co., 1970.

Galana, Laurel. "Radical reproduction: X without Y." In *The Lesbian Reader*. Edited by Gina and Laurel. Oakland, California: Amazon Press, 1975.

Gaskin, I. M. *Spiritual Midwifery*, rev. ed. Summertown, Tenn.: Book Publishing Co., 1978.

Gilgoff, A. *Home Birth*. New York: Coward, McCann and Geoghegan, 1978.

Gordon, R. E., E. Kapostins and K. K. Gordon. "Factors in postpartum emotional adjustment." *Obstetrics and Gynecology* 25 (1965): 158–166.

Gordon, R. E., and K. K. Gordon. "Social factors in the prediction and treatment of emotional disorders in pregnancy." *American Journal of Obstetrics and Gynecology* 77 (1959): 1074–1083.

Gregory, Samuel. *Man Midwifery Exposed and Corrected*. Boston, 1848.

Hanford, Jean M. "Pregnancy as a state of conflict." *Psychological Reports* 22 (1968): 1313–1342.

Hausknecht, R., and J. H. Heilman. *Having a Caesarean Baby.* New York: E. P. Dutton, 1978.

Hazell, L. D. *Commonsense Childbirth.* New York: Berkely Publishing Co., 1969.

Home Oriented Birth Experience [H.O.M.E.]. *A Comprehensive Guide to Home Birth.* Available from 511 New York Ave., Takoma Park, Washington, D.C., 1976.

Jex-Blake, Sophia. "Women as practitioners of midwifery." *Lancet* 2 (1870).

Kippley, Sheila. *Breast Feeding and Natural Child Spacing: The Ecology of Natural Mothering.* Middlesex, England: Penguin Books, 1974.

Kitzinger, Sheila. *The Experience of Childbirth.* Middlesex, England: Penguin Books, 1973, 1977.

————. *Giving Birth: The Parents' Emotions in Childbirth.* New York: Schocken Books, 1977.

————, and J. A. Davis. *The Place of Birth.* Oxford: University Press, 1978.

Kramer, R. *Giving Birth: Childbearing in America Today.* Chicago: Contemporary Books, 1978.

Krobin, Francis E. "The American midwife controversy: A crisis of professionalization." *Bulletin of the History of Medicine* 40 (1966).

Lamaze, Fernand. *Painless Childbirth: The Lamaze Method.* New York: Pocket Books, 1955.

Lazare, Jane. *The Mother Knot.* New York: Dell Publishing Co., 1976.

Leboyer, Frederick. *Birth Without Violence.* New York: Knopf, 1975.

Lewi, Maurice J. "What shall be done with the professional midwife." *Transactions of the Medical Society of the State of New York* (1902).

Long, Raven, ed. *Birth Book.* Felton, California: Genesis Press, 1972.

Lubin, Bernard, Sprague H. Gardener, and Aleda Roth. "Mood and somatic symptoms during pregnancy." *Psychosomatic Medicine* 37 (1975).

McBride, Angela Barron. *The Growth and Development of Mothers.* New York: Perennial Library, Harper and Row, 1973.

McCauley, Carole Spearin. *Pregnancy After 35.* New York: E. P. Dutton and Co., 1976.

McFarlane, Aida. *The Psychology of Childbirth.* Cambridge, Mass.: Harvard University Press, 1977.

Medvin, Jeanne O'Brien. *Prenatal Yoga and Natural Childbirth.* Albion, California: Freestone Publishing Co., 1974.

Mehl, Lewis E. "Options in maternity care." *Women and Health* 2 (Sept./Oct., 1977): 29–42.

Mengert, William F. "The origin of the male midwife." *Annals of Medical History* 4 (1932).

Milinaire, Caterine. *Birth*. New York: Harmony Books, 1974.

Movland, Egbert. *Alice and the Stork: Or the Rise in the Status of the Midwife as Exemplified in the Life of Alice Gregory, 1867–1944*. London: Hodder and Stoughton, 1951.

Noble, Elizabeth. *Essential Exercises for the Childbearing Years*. Boston: Houghton Mifflin, Co., 1976.

Pryor, Karen. *Nursing Your Baby*. New York: Pocket Books, 1963.

Radcliff, Walter. *Milestones in Midwifery*. Bristol, England: John Wright, 1967.

*Rich, Adrienne. *Of Woman Born: Motherhood as Experience and Institution*. New York: Norton, 1976.

*Rose, Hilary, and Jalna Hanmer. "Women's liberation and the technological fix." In *The Political Economy of Science: Ideology of/in the Natural Sciences*. Edited by Hilary Rose and S. Rose. London: The Macmillan Press, 1976. Reprinted in *Ideology of/in the Natural Sciences*. Hilary Rose and S. Rose, eds. Boston, Mass.: G. K. Hall & Co.; Cambridge, Mass.: Schenkman Publishing Co., forthcoming 1979.

Rosengren, W. R. "Social sources of pregnancy as illness or normality." *Social Forces* 39 (March 1961): 260–267.

Sablosky, A. H. "The power of the forceps: A comparative analysis of the midwife—historically and today." *Women and Health* 1 (Jan./Feb., 1976): 10–13.

Shainess, Natalie: "Psychological problems associated with motherhood." In *American Handbook of Psychiatry,* vol. III. Edited by Silvano Arieti. New York: Basic Books, 1966.

Shaw, N. S. *Forced Labor: Maternity Care in the United States*. New York: Pergamon Press, 1974.

Sousa, Marion. *Childbirth at Home*. New York: Bantam Books, 1976.

Stewart, D., and L. Stewart. eds. *Safe Alternatives in Childbirth*. Chapel Hill, North Carolina: NAPSAC, 1976.

Tanzer, Debora. *Why Natural Childbirth*. New York: Schocken Books, 1972.

Tucker, Tarvez, with Elizabeth Bing. *Prepared Childbirth*. New Canaan, Conn.: Tobey Publishing Co., 1975.

Ward, C., and F. Ward. *The Home Birth Book*. Garden City, N.Y.: Dolphin Books, 1977.

Wertz, R. W., and D. C. Wertz. *Lying-In: A History of Childbirth in America*. New York: The Free Press, 1977.

Willoghby, Percival. *Observations in Midwifery*. Edited by Henry Blenkinsop. Yorkshire, England: S. R. Publishers, 1863.

Woolfolk, W., and J. Woolfolk. *The Great American Birth Rite: Babies as Big Business.* New York: Dial Press, 1975.

Wright, Erma. *The New Childbirth.* New York: Pocket Books, 1976.


LESBIAN HEALTH

Chico Feminist Women's Health Center. *Health Care for Lesbians.* Chico Feminist Women's Health Center, 330 Flume St., Chico, California 95926.

Fenwick, R. D. *The Advocate Guide to Gay Health.* New York: E. P. Dutton, 1978.

Hornstein, Frances. *Lesbian Health Care.* OFWHC, 2930 McClure Street, Oakland, California 94609.

Radicalesbians Health Collective. "Lesbians and the health care system." In *Out of the Closets: Voices of Gay Liberation.* Edited by Darla Jan and Allen Young. Moonachie, N. J.: Pyramid Publications, 1974.

*Santa Cruz Women's Health Center. *Annotated Bibliography on Lesbian Health Issues.* Santa Cruz Women's Health Center, 250 Locust Street, Santa Cruz, California 95060.

Wysor, Betty. *The Lesbian Myth.* New York: Random House, 1975.


PSYCHOLOGY

APA Task Force on Sex Bias and Sex-Role Stereotyping in Psychotherapeutic Practice. "Report of the Task Force." *American Psychologist* (Dec. 1975): 1169–1175.

Aslin, Alice L. "Feminist and community mental health center psychotherapists: Expectations of mental health for women." *Sex Roles* 3 (Dec. 1977).

Bardwick, Judith M. *Feminine Personality and Conflict.* Monterey, California: Brooks-Cole, 1970.

————. *Psychology of Women: A Study of Bio-Cultural Conflicts.* New York: Harper and Row, 1971.

————, ed. *Readings on the Psychology of Women.* New York: Harper and Row, 1972.

Bart, Pauline. "Depression in middle-aged women." In *Women in a Sexist Society.* Edited by Vivian Gornick and Barbara Moran. New York: Basic Books, 1971.

Bem, S. L. "Sex-role adaptability: one consequence of psychological androgyny." *Journal of Personality and Social Psychology* 31 (1975): 634–643.

Bonaparte, Marie. "Passivity, masochism and feminity." *International Journal of Psychoanalysis* 16 (1935): 235–333.

Bosma, Barbara J. "Attitudes of women therapists toward women clients, or a comparative study of feminist therapy." *Smith College Studies in Social Work* 46 (Nov. 1975): 1.

*Broverman, Inge K., et al. "Sex role stereotypes and clinical judgements of mental health." *Journal of Consulting and Clinical Psychology* 34 (Feb. 1970): 1–7.

Brown, Judith. "Feminism and its implications for therapy." *Radical Therapist* 1 (1970): 5–6.

Brown, Phil, ed. *Radical Psychology.* New York: Colophon Books, Harper and Row, 1973.

Bruch, Hilde. *The Golden Cage: The Enigma of Anorexia Nervosa.* Cambridge, Mass.: Harvard University Press, 1978.

Castillejo, Claremont. *Knowing Woman: A Feminine Psychology.* New York: Colophon Books, Harper and Row, 1973.

Chapman, Joseph Dudley. *The Feminine Mind and Body: The Psycho-sexual and Psychosomatic Reactions of Woman.* New York: Philosophical Library, 1967.

Chesler, Phyllis. *About Men: A Psycho-Sexual Meditation.* New York: Simon and Schuster, 1978.

*————. *Women and Madness.* New York: Doubleday, 1972.

*Chodorow, N. *The Reproduction of Mothering: Psychoanalysis and the Sociology of Gender.* Berkeley: University of California Press, 1978.

Clancey, K., and W. Gove. "Sex differences in mental illness." *American Journal of Sociology* 80 (1974): 204–216.

Coie, J. D., B. F. Pennington, and H. H. Buckley. "Effects of situational stress and sex roles on the attribution of psychological disorder." *Journal of Consulting and Clinical Psychology* 4 (1974), 559–568.

Cox, Sue, ed. *Female Psychology: The Emerging Self.* Chicago: Science Research Associates, 1976.

Deutsch, Helene. *The Psychology of Women,* vols. I and II. New York: Grune and Stratton, 1944.

*Dinnerstein, Dorothy. *The Mermaid and the Minotaur.* New York: Harper/Colophon Books, 1976.

Dohrenwend, B. P., and B. S. Dohrenwend. "Sex differences and Psychiatric disorders." *American Journal of Sociology* (May 1976): 1447–1454.

Donelson, E., and J. E. Gullahorn. *Women: A Psychological Perspective.* New York: John Wiley and Sons, 1977.

Druss, Vicki, and Mary Sue Henifin. "Why Are So Many Anorexics Women?" In *Women Look at Biology Looking at Women.* R. Hubbard, M. S. Henifin, and B. Fried, eds. Boston, Mass.: G. K. Hall & Co.; Cambridge, Mass.: Schenkman Publishing Co., 1979.

Franks, V., and W. Burtle. *Women in Therapy: New Psychotherapies for a Changing Society.* New York: Brunner/Mazel Publishers, 1974.

Freud, Sigmund. "Female sexuality." In *Collected Papers,* vol. 5. London: Hogarth, 1950.

————. "Feminity." In *New Introductory Lectures on Psycho-analysis.* Edited by J. Strachey. New York: W. W. Norton, 1965 (originally 1933), pp. 112–135.

————. "The psychology of women: biology as destiny." In *Women in a Made-Made World.* Edited by Nona Glazer-Malbin and Helen Youngelson Waehrer. Chicago: Rand McNally, 1972, pp. 58–61.

————. "Some psychical consequences of the anatomical distinction between the sexes." In *Women and Analysis.* Edited by Jean Strauss. New York: Dell Publishing Co., 1974.

————. *Three Essays on the Theory of Sexuality.* Translated and edited by James Strachey. New York: Basic Books, 1963.

————, and J. Breuer. "Studies in hysteria." In *The Complete Freud,* vol. II. London: Hogarth Press, 1955.

Friedan, Betty. "The sexual solipsism of Sigmund Freud." In *The Feminine Mystique.* New York: Dell, 1963.

Goldman, George D., and Donald Milman, eds. *Modern Woman: Her Psychology and Sexuality.* Springfield, Illinois: C. C. Thomas, 1969.

Gove, Walter R. "Adult sex roles and mental illness." *American Journal of Sociology* 78 (Jan. 1973): 812–835.

Gove, William. "The relationship between sex roles, marital status and mental illness." *Social Forces* 51 (1972): 34–44.

Greek, Frances E. "A serendipitous finding: Sex roles and schizophrenia." *Journal of Abnormal and Social Psychology* 69 (1964): 392–400.

Gump, Janice Porter. "Sex role attitudes and psychological well-being." *Journal of Social Issues* 29 (1973): 79–92.

H. D. [Hilda Doolittle] *Tribute to Freud.* New York: McGraw Hill, 1974 (originally 1944).

Hall, C. "A modest confirmation of Freud's theory of distinction between the superego of men and women." *Journal of Abnormal and Social Psychology* 69 (1964): 440–442.

Hansson, Laura. *The Psychology of Woman.* London: G. Richards, 1899.

Harding, Mary Esther. *Woman's Mysteries: Ancient and Modern.* New York: G. P. Putnam's, 1971.

————. *The Way of All Women: A Psychological Interpretation.* New York: Harper/Colophon Books, 1970.

Hays, H. R. *The Dangerous Sex: Thy Myth of Feminine Evil.* New York: Putnam, 1964.

Hinkle, Beatrice. "On the arbitrary use of the terms 'masculine' and 'feminine.' " *Psychoanalytic Review* 7 (1920): 15–30.

Horner, Matina. "The motive to avoid success." *Psychology Today* 36 (1969).

————. "Toward an understanding of achievement related conflicts in women." *Journal of Social Issues* 28 (1972): 157–176.

*Horney, Karen. *Feminine Psychology*. New York: W. W. Norton and Company, 1967.

Hyde, J.S., and B. G. Rosenberg. *Half the Human Experience: The Psychology of Women*. Lexington, Mass.: D. C. Heath, 1976.

Jung, Carl G. "Psychological aspects of the mother archetype." In *Four Archetypes*. Princeton: Bollingen, 1973.

*Kaplan, Alexandra, and Joan P. Bean, eds. *Beyond Sex-Role Stereotypes: Readings Toward a Psychology of Androgyny*. Boston: Little Brown and Co., 1976.

Lennane, K. Jean., and R. J. Lennane. "Alleged psychogenic disorders in women—a possible manifestation of sexual prejudice." *New England Journal of Medicine* 288 (1973): 288–292.

Levine, Saul, Louise Karin, and Eleanor Lee Levine. "Sexism and psychiatry." *American Journal of Orthopsychiatry* 44 (April 1974).

Levy, R. Psychosomatic symptoms and women's protest: Two types of reaction to structural strain in the family. *Journal of Health and Social Behavior* 17 (1976): 121–133.

Manalis, Sylvia A. "The psychoanalytic concept of feminine passivity: A comparative study of psychoanalytic and feminist views." *Comprehensive Psychiatry* 17 (1976).

Mander, Anica Vesel, and Anne Kent Rush. *Feminism as Therapy*. New York: Random House, 1974.

Miller, Jean Baker, ed. *Psychoanalysis and Women*. New York: Penguin Books, 1973.

————. *Toward a New Psychology of Women*. Boston: Beacon Press, 1976.

*Millett, Kate. "Freud and the influence of psychoanalytic thought." In *Sexual Politics*. New York: Doubleday, 1970, pp. 176–233.

*Mitchell, Juliet. *Psychoanalysis and Feminism*. New York: Pantheon Books, 1974.

Nelson, Marie Coleman, and Jean Ikenberry eds. *Psychosexual Imperatives: Their Roles in Identity Formation*. New York: Human Sciences Press, 1978.

Neumann, Erich. *Amor and Psyche: The Psychic Development of the Feminine*. Princeton: Princeton University Press, 1956.

Patrick, G.T.W. "The psychology of woman." *Popular Science Monthly* 47 (1895): 209–224.

Radloff, Lenore. "Sex differences in depression." *Sex Roles* 1 (Sept. 1975): 249–265.

Rieff, Philip. *Freud: The Mind of the Moralist*. New York: Doubleday, 1959. [See Chapter V: Sexuality and Domination.]

*Seiden, Anne M. "Overview: Research on the psychology of women. I. Gender differences and sexual and reproductive life." *American Journal of Psychiatry* 133 (1976): 995–1007.

*———— "Overview: Research on the psychology of women. II. Women

in families, work, and psychotherapy." *American Journal of Psychiatry* 133 (1976): 1111–1123.

Sherman, Julia A. *On the Psychology of Women: A Survey of Empirical Studies.* Springfield, Illinois: Charles C. Thomas, 1971.

Smith, Dorothy E., and Sara J. David, eds. *Women Look at Psychiatry.* Vancouver, B.C.: Press Gang Publishers, 1975.

Smith-Rosenberg, Carroll. "The hysterical woman: Sex roles and role conflict in nineteenth-century America." *Social Research* 39 (1972).

Stein, Robert. "Phallos and feminine psychology." In *Incest and Human Love: The Betrayal of the Soul in Psychotherapy.* Edited by Robert Stein. Baltimore: Penguin, 1973.

Strauss, Jean, ed. *Women and Analysis: Dialogues on Psychoanalytic Views of Femininity.* New York: Dell Publishing Co., 1974.

Thompson, Clara. "Cultural pressures in the psychology of women." *Psychiatry* 5 (1942): 331–339.

Walstedt, Joyce Jennings. *The Psychology of Women: A Partially Annotated Bibliography.* Pittsburgh: KNOW, Inc., 1972.

*Weisstein, Naomi. *Kinder, Kuche, Kirche as Scientific Law: Psychology Constructs the Female.* Boston: New England Free Press, 1968.

*––––––. "Psychology constructs the female; or the fantasy life of the male psychologist (with some attention to the fantasies of his friends, the male biologist and the male anthropologist)." *Social Education* 35 (1971): 362–373.

Wesley, Carol. "The women's movement and psychotherapy." *Social Work* 20 (March 1975): 120–125.

Williams, Elizabeth Friar. *Notes of a Feminist Therapist.* New York: Dell Publishing Co., 1976.

Williams, Juanita H. *Psychology of Women.* New York: Norton, 1977.

Women and Therapy Collective. *Off the Couch: A Woman's Guide to Psychotherapy.* Cambridge, Mass.: Goddard Cambridge Graduate Program in Social Change, 1976.

Zeldow, Peter. "Clinical judgement: A search for sex differences." *Psychological Reports* 37 (1975): 1135–1142.

––––––. "Psychological androgyny and attitudes towards feminism." *Journal of Consulting and Clinical Psychology* 44 (Feb. 1976): 1.

## SEXUALITY

*Abott, Sidney, and Barbara Love. *Sappho Was a Right-on Woman: A Liberated View of Lesbianism.* New York: Stein and Day, 1972.

Adams, C., and R. Laurikietis, *The Gender Trap: Book 2—Sex and Marriage.* London: Virago Books, 1976.

Barbach, Lonnie Garfield. *For Yourself: The Fulfillment of Female Sexuality.* New York: A Signet Book, New American Library, 1975.

————. "Group treatment for pre-orgasmic women." *Journal of Sex and Marital Therapy* (1975): 2.

————, and Toni Ayres. "Group process for women with orgasmic difficulties." *Personnel and Guidance Journal* 54 (March 1976): 389–391.

Barnett, M.C. "Vaginal awareness in the infancy and childhood of girls." *Journal of the American Psychoanalytic Association* 14 (1960): 129–141.

Bell, Alan P., and Martin S. Weinberg. *Homosexualities: A Study of Diversity Among Men and Women.* New York: Simon and Schuster, 1978.

Belliveau, Fred, and Lin Richter. *Understanding Human Sexual Inadequacy.* New York: Bantam Books, 1970.

Benedek, Therese. *Psychosexual Function in Women.* New York: Ronald Press, 1952.

Benedek, T. "Sexual functions in women and their disturbance." *American Handbook of Psychiatry,* Vol. 1. Edited by S. Arietta. New York: Basic Books, 1959.

————, and B. Rubenstein. "The sexual cycle in women: the relationship between ovarian function and psychodynamic processes." *Psychosomatic Medicine* 1 (1935): 246–270.

Berghe, Pierre van den. *Age and Sex in Human Society: A Biosocial Perspective.* Belmont, California: Wadsworth, 1973.

Bergler, E., and W.S. Kroger. *Kinsey's Myth of Female Sexuality.* New York: Grune and Stratton, 1954.

Blank, J., and H. L. Cottrell. *I Am My Lover.* Burlingame, California: Down There Press, 1978.

Bonaparte, Marie. *Female Sexuality.* New York: International University Press, 1956.

Brown, D. "Female orgasm and sexual inadequacy." In *An Analysis of Human Sexual Response.* Edited by E. Brecher. New York: New American Library, 1966.

Chassequet-Smirgel, J. *Female Sexuality: New Psychoanalytic Views.* Ann Arbor: University of Michigan Press, 1970.

Christenson, C.V., and J.H. Gagnon. "Sexual behavior in a group of older women." *Journal of Gerontology* 20 (1965): 351–356.

*Country Women* [issue on sexuality]. Issue No. 15 (April 1975). Available from Box 51, Albion, California 95410.

Dickinson, R. L., and H. H. Pierson. "The average sex life of American women." *Journal of the American Medical Association* 85 (1925): 1113–1117.

Dodsen, Betty. *Liberating Masturbation.* Available from Box 1933, New York, N. Y. 10001, 1974.

Dworkin, Andrea. *Woman Hating: A Radical Look at Sexuality.* New York: Dutton, 1976.

Elkan, E. "Evolution of female orgastic ability—a biologic survey." *International Journal of Sexology* 2 (1948): 84–93.

Ellis, Havelock. *The Erotic Rights of Women*. London: British Society for the Study of Sex Psychology, 1918.

Ellis, Albert. "Is the vaginal orgasm a myth?" In *Sex, Society and the Individual*. Edited by A. P. Pillay and A. Ellis. Bombay: International Journal of Sexology Press, 1953.

Feinbloom, Deborah Heller. *Transvestites and Transsexuals*. New York: Dell Publishing Co., 1976.

Fisher, S. *The Female Orgasm: Psychology, Physiology, Fantasy*. New York: Basic Books, 1973.

————. *Understanding the Female Orgasm*. New York: Bantam, 1973.

*Haller, John S., and Robin M. Haller. *The Physician and Sexuality in Victorian America*. New York: W. W. Norton and Co., 1974.

*Hammer, Signe, ed. *Women: Body and Culture—Essays on the Sexuality of Women in a Changing Society*. New York: Perennial Library, Harper and Row, 1974.

Heiman, J., L. LoPiccolo, and J. LoPiccolo. *Becoming Orgasmic: A Sexual Growth Program for Women*. New York: Prentice Hall, 1976.

*Hite, Shere. *The Hite Report: A Nationwide Study of Female Sexuality*. New York: Dell Publishing Co., 1976.

————. *Sexual Honesty*. New York: Warner Books, 1974.

Horney, K. "The denial of the vagina: A contribution to the problem of genital anxieties specific to women." *International Journal of Psychoanalysis* 14 (1933): 55–70.

Kane, Francis J., Morris A. Lipton, and John A. Ewing. "Hormonal influences in female sexual response." *Archives of General Psychiatry* 20 (1969): 202–209.

Kelly, G. L. *Sexual Feeling in Woman*. Augusta, Georgia: Elkay Press, 1930.

*Kinsey, A. C., et al. *Sexual Behavior in the Human Female*. Philadelphia: W. B. Saunders and Co., 1953.

*————— et al. *Sexual Behavior in the Human Male*. Philadelphia: W. B. Saunders and Co., 1948.

Klaich Dolores. *Woman + Woman: Attitudes toward Lesbianism*. New York: William Morrow and Co., 1974.

Kline-Graber, Georgia, and Benjamin Graber. *Woman's Orgasm*. New York: Popular Library, 1976.

Koedt, Anne. "The myth of the vaginal orgasm." In *Liberation Now*. Edited by Babcor and Belkin. New York: Dell, 1971, pp. 311–320.

Kronhausen, Phyllis, and Eberhard Kronhausen. *The Sexually Responsive Woman*. New York: Ballantine, 1965.

Loewenstein, Sophie. "An overview of some aspects of female sexuality." *Social Casework* 59 (1978): 106–115.

LoPiccolo, J., and M. A. Lobitz. "The role of masturbation in the treatment of sexual dysfunction." *Archives of Sexual Behavior* 2 (1972): 163–171.

Lorand, S. "Contributions to the problem of vaginal orgasm." *International Journal of Psychoanalysis* 20 (1939): 432–438.

McDermott, Sandra. *Female Sexuality: Its Nature and Conflict.* First Things First, 2334 Ontario Road, N.W., Washington, D.C. 20009.

*Martin, Del, and Phyllis Lyon. *Lesbian/Woman.* New York: Bantam Books, 1972.

Maslow, A. H., H. Rand, and S. Newman. "Some parallels between sexual and dominance behavior of infrahuman primates and the fantasies of patients in psychotherapy." *Journal of Nervous and Mental Diseases* 313 (1960): 202–212.

Masters, William H., and Virginia Johnson. *Human Sexual Inadequacy.* Boston: Little Brown and Co., 1970.

*——————. *Human Sexual Response.* Boston: Little Brown and Co., 1966.

Maxwell, R. J. "Quiz: Female sexuality in primitive cultures." *Medical Aspects of Human Sexuality* 1 (Jan. 1973).

Parsons, Elsie Clews. *The Old-fashioned Woman: Primitive Fancies About Sex.* New York: G. P. Putnam, 1913.

Ploss, Herman Heinrich. *Woman in the Sexual Relation: An Anthropological and Historical Survey.* New York: Medical Press of New York, 1964.

Ponse, Barbara. *Identities in the Lesbian World: The Social Construction of Self.* Westport, Conn.: Greenwood Press, 1978.

Raymond, Janice G. "Transsexualism: The ultimate homage to sex-role power." *Chrysalis* 3 (1978): 11–23.

——————. *The Transsexual Empire: The Making of the She-Male.* Boston: Beacon Press, 1979.

Robinson, W. J. *Woman: Her Sex and Love Life.* New York: Eugenics Publishing Co., 1929.

Rush, Anne Kent. *Getting Clear: Body Work for Women.* New York: Random House, 1973.

*Seaman, Barbara. *Free and Female: The New Sexual Role of Women.* Greenwich, Conn.: Fawcett Crest Book, 1972.

Seidenberg, Robert. "Is sex without sexism possible?" *Sexual Behavior* 46 (1972).

Sherfey, Mary Jane. "Female sexuality and psychoanalytic theory." In *Woman in a Man-Made World.* 2nd Edition. Edited by Nona Glazer and Helen Waehrer. Chicago: Rand McNally and Co., 1972.

*——————. *The Nature and Evolution of Female Sexuality.* New York: Random House, 1966.

*Singer, June. *Androgyny: Toward a New Theory of Sexuality.* Garden City, New York: Anchor Books, 1977.

Sisley, Emily L., and Bertha Harris. *The Joy of Lesbian Sex.* New York: Crown Publishers, 1977.

Smart, C., and B. Smart. *Women, Sexuality, and Social Control*. London: Routledge and Kegan Paul, 1978.

Stekel, Wilhelm. *Frigidity in Women*, 2 volumes. London: Boni and Liveright, 1926.

Teeters, Kass. *Female Sexuality*. Women Inc., San Jose, California. 95132.

Wallace, F. *Masturbation: A Woman's Handbook*. Bloomfield, N.J.: R. J. Williams Publishers, 1975.

Weeks, Jeffrey. *Coming Out: Homosexual Politics in Britain from the Nineteenth Century to the Present*. London: Quartet Books; New York: Horizon Press, 1977.

Wysor, Betty. *The Lesbian Myth*. New York: Random House, 1974.


RAPE AND INCEST


Armstrong, L. *Kiss Daddy Goodnight: Speak Out on Incest*. New York: Hawthorn Books, 1978.

Barnes, D. L. *Rape: A Bibliography 1965–1975*. Troy, N. Y.: Whiteson Publishers, 1977.

*Brownmiller, Susan. *Against Our Will: Men, Women, and Rape*. New York: Bantam Books, 1975.

Burgess, A., and L. Holmstrom. *Rape: Victims of Crisis*. Bowie, Maryland: Brady (Prentice-Hall), 1974.

Butler, S. *Conspiracy of Silence: The Trauma of Incest*. San Francisco: New Glide Publications, 1978.

Chappell, D., R. Geis, and G. Geis, eds. *Forcible Rape: The Crime, the Victim, and the Offender*. New York: Columbia University Press, 1977.

Clark, L., and D. Lewis. *Rape: The Price of Coercive Sexuality*. Toronto, Canada: The Women's Press, 1977.

Connell, Noreen, and Cassandra Wilson. *Rape: The First Source Book for Women*. New York: A Plume Book, New American Library, 1974.

Forward, S., and C. Buck. *Betrayal of Innocence: Incest and its Devastation*. Los Angeles: J. P. Tarcher, 1978.

Gager, Nancy, and Kathleen Schurr. *Sexual Assault: Confronting Rape in America*. New York: Grosset and Dunlap, 1976.

Herman, Judith, and Lisa Hirschman. "Incest between fathers and daughters." *The Sciences* 17 (Nov. 1977): 4–7.

Hilberman, E. *The Rape Victim*. New York: Basic Books, 1976.

Horos, C. V. *Rape*. New Canaan, Conn.: Tobey Publishing Co., 1974.

Medea, Andra, and Kathleen Thompson. *Against Rape*. New York: Farrar, Straus, and Giroux, 1974.

Russell, Diana E. H. *The Politics of Rape: The Victim's Perspective*. New York: Stein and Day, 1976.

Russell, D.E.H., and N. Van de Ven. *The Proceedings of the International Tribunal on Crimes Against Women*. Millbrae, California: Les Femmes, 1976.

St. Louis Feminist Research Project. *The Rape Bibliography: A Collection of Abstracts*. St. Louis, Missouri: Edy Netter, 1976.


CANCER


Campion, R. *The Invisible Worm* [alternatives to breast surgery]. New York: Avon Books, 1975.

*Cope, Oliver. *The Breast: Its Problems, Benign and Malignant, and How to Deal With Them*. Boston: Houghton Mifflin, 1977.

Coweles, J. *Informed Consent* [breast cancer]. New York: Coward, McCann and Geoghegan, 1976.

*Crile, George, Jr. *What Women Should Know About the Breast Cancer Controversy*. New York: Pocket Books, 1973.

Greenfield, N. S. *First Do No Harm: A Dying Woman's Battle Against the Physicians and the Drug Companies Who Misled Her About the Pill*. New York: Sun River Press, 1976.

Jameson, DeeDee, and Roberta Schwalb. *Hysterectomy: Taking Charge of Your Own Body*. Englewood Cliffs, N. J.: Spectrum Books, 1978.

*Kushner, Rose. *Why Me: What Every Woman Should Know About Breast Cancer to Save Her Life*. New York: A Signet Book, New American Library, 1977.

Newman, J. H. *What Every Woman Should Know About Breast Cancer*. Canoga Park, California: Major Books, 1976.

Nugent, N. *Hysterectomy*. Garden City, N.Y.: Doubleday and Co., 1972.

*Seaman, Barbara, and Gideon Seaman. *Women and the Crisis in Sex Hormones*. New York: Rawson Associates, 1977.

Seaman, S. S. *Always A Woman: What Every Woman Should Know About Breast Surgery*. Larchmont, N.Y.: Argonaut Books, 1965.

Smith, Donald C., et al. "Association of exogenous estrogen and endrometrial carcinoma." *New England Journal of Medicine* 293 (Dec. 1975): 1164–1167.

Strax, P. *Early Detection: Breast Cancer is Curable*. New York: New American Library, 1974.

Weiss, Kay. "Vaginal cancer: An iatrogenic disease?" *International Journal of Health Services* 5 (1975): 235–252.

Zalon, J., with J. L. Block. *I Am Whole Again: A Case for Reconstruction After Breast Surgery*. New York: Random House, 1978.

Ziel, Harry K., and William D. Finkle. "Increased risk of endometrial carcinoma among users of conjugated estrogens." *New England Journal of Medicine* 293 (Dec. 1975): 1167–1170.

WOMEN AND EXERCISE

American Alliance for Health, Physical Education and Recreation. *Women's Athletics: Coping With Controversy*. Washington, D.C.: AAHPER, 1974.

Balazs, Eva. *In Quest of Excellence: A Psycho-social Study of Female Olympic Champions*. Warwick, N.J.: Hoctor Products for Education, 1975.

Barilleaux, D., and J. Murray. *Inside Weight Training for Women*. New York: Contemporary, 1978.

Beall, Elizabeth. "The Relation of Various Anthropometric Measurements of Selected College Women to Success in Certain Physical Activities." Master's Thesis, Columbia University, 1939.

Butt, D. S. *Psychology of Sport*. New York: Van Nostrand Reinhold Co., 1976. [See Chapter 4: Sex Roles in Sport.]

Garrick, J. G., and R. K. Regman. "Girls' sports injuries in high school athletics." *Journal of the American Medical Association* 239 (1978): 2245–2248.

Gerber, E. W., et al. *The American Woman in Sport*. Menlo Park, California: Addison-Wesley Publishing Co., 1974.

Haycock, C. E., and Gillette, J. "Susceptibility of women athletes to injury: Myth vs. reality." *Journal of the American Medical Association* 236 (1976): 163–165.

Jacobs, K. F. *Girlsports*. New York: Bantam Books, 1978.

Lance, Kathleen. *Getting Strong: A Woman's Guide to Realizing Her Physical Potential*. Indianapolis, Indiana: Bobbs Merrill Co., 1978.

Oglesby, C. A. *Women and Sport: From Myth to Reality*. Philadelphia: Lea and Febiger, 1978.

Runner's World Magazine. *The Female Runner*. Mountain View, California: World Publications, 1974.

Ullyot, Joan. *Women's Running*. Mountain View, California: World View Publications, 1976.


BIOGRAPHY

Baker, Rachel. *The First Woman Doctor: The Story of Elizabeth Blackwell, M.D.* New York: Julian Messner, 1944.

Balfour, Margaret Ida. *The Work of Medical Women in India*. London and New York: Oxford University Press, 1929.

Barringer, Emily A. *Bowery to Bellevue: The Story of New York's First Woman Ambulance Surgeon*. New York: W. W. Norton, Co., 1950.

Bass, Elizabeth. "Dispensaries founded by women physicians in the Southland." *Journal of the American Medical Women's Association* 2 (1947).

————. "Pioneer women doctors in the South." *Journal of the American Medical Women's Association* 2 (1947).

Bell, Enid Mobely. *Storming the Citadel: The Rise of the Woman Doctor*. London: Constable, 1953.

Bennett, A. H. *English Medical Women: Glimpses of Their Work in Peace and War*. London: Pitman, 1915.

Bluemel, Elinor. *Florence Sabin: Colorado Woman of the Century*. Boulder: University of Colorado Press, 1957.

Bowen, Gertrude Maude. *I Have Lived* [autobiography of a doctor]. Grantham: Stanborough Press, 1973.

Breckenridge, Mary. *Wide Neighbors: A Story of the Frontier Nursing Service*. New York: Harper and Bros. 1952.

Campbell, J. Menzies. "Bygone women dentists." *Journal of the Canadian Dental Association* (April 1948).

Dally, Ann G. *Cicely, The Story of a Doctor*. London: Victor Gollancz, 1968.

Douglass, Emily Taft. *Margaret Sanger: Pioneer of the Future*. New York: Holt, Rinehart, and Winston, 1970.

Dykeman, Wilma. *Too Many People, Too Little Love: Edna Rankin McKinnon, Pioneer for Birth Control*. New York: Holt, 1974.

Edwards, R. W. "The first woman dentist: Lucy Hobbs Taylor, D.D.S. 1833–1910." *Bulletin of the History of Medicine* 25 (1915): 277–283.

Elia, Joseph J. "Alice Hamilton, 1869–1970." *New England Journal of Medicine* 283 (1970).

Emerson, Gladys A. "Agnes Fay Morgan and early nutrition discoveries in California." *Federation Proceedings* 36 (May 1977): 1911–1914.

Fleming, Alice. *Doctors in Petticoats*. Lippincott, 1964.

Grant, Madeleine P. *Alice Hamilton, Pioneer Doctor in Industrial Medicine*. New York and London: Abelard-Schuman, 1967.

Hamilton, Alice. *Exploring the Dangerous Trades: Autobiography of an Industrial Toxicologist with Jane Addams at Hull House*. Boston, 1943.

————. "Pioneering in industrial medicine." *Journal of the American Medical Women's Association* 2 (1947).

Hays, Elinor Rice. *Those Extraordinary Blackwells: The Story of a Journey to a Better World*. New York: Harcourt, Brace and World, 1967.

Hughes, Muriel Joy. *Women Healers in Medieval Life and Literature*. Freeport, New York: Books for Libraries Press, 1968 (originally 1943).

Hume, Ruth Fox. *Great Women of Medicine*. New York: Random House, 1964.

Jacobi, Mary Putnam. *Life and Letters of Mary Putnam Jacobi*. Edited by Ruth Putnam. New York: G. P. Putnam's Sons, 1925.

——————. *Mary Putnam Jacobi, M.D.: A Pathfinder in Medicine, With Selections from Her Writings and a Complete Bibliography*. Edited by the Women's Medical Association of New York City. New York: G. P. Putnam's Sons, 1928.

——————. "Women in medicine." In *Women's Work in America*. Edited by Annie Nathan Meyer. New York: Henry Holt, 1891.

Johnston, Malcolm Sanders. *Elizabeth Blackwell and Her Alma Mater: The Story of the Documents*. Geneva, N.Y.: W. F. Humphrey Press, 1947.

King-Salmon, Frances W. *House of a Thousand Babies: Experiences of an American Woman Physician in China 1922–1940*. New York: Exposition Press, 1968.

Knapp, Sally. *Women Doctors Today*. New York: Thomas Y. Crowell, Co., 1947.

Lovejoy, Esther. *Women Doctors of the World*. New York: Macmillan, 1957.

Lutzker, Edythe. *Edith Peckey-Phipson, M.D: The Story of England's Primary Woman Doctor*. New York: Exposition Press, 1973.

MacDermot, H. E. *Maude Abbott: A Memoir*. Toronto, Macmillan, 1941.

McFerran, Ann. *Elizabeth Blackwell, First Woman Doctor*. New York: Grosset and Dunlap, 1966.

McMaster, Gilbert Totten. "The first woman practitioner of midwifery and the care of infants in Athens, 300 B.C. [Agnodice]." *American Medicine* 18 (1912): 202–205.

Manson, Cecil, and Celia Manson. *Doctor Agnes Bennett*. London: Michael Joseph, 1960.

Manton, J. *Elizabeth Garrett Anderson* [first female English physician]. New York: Dutton, 1965.

Marks, Geoffrey, and William K. Beatty. *Women in White*. New York: Scribner's 1972.

Morton, Rosalie Slaughter. *A Woman Surgeon: The Life and Work of Rosalie Slaughter Morton*. New York: Stokes, 1937.

Noble, Ins. *First Woman Ambulance Surgeon: Emily Barringer*. New York: Julian Messner, 1962.

Overholser, Winfred. "Dorothea Lynde Dix: A note." *Bulletin of the History of Medicine* 9 (1941).

Pearsall, Ronald. "A pioneer bone setter: Mrs. Sarah Mapp." *Practitioner* 195 (1965).

Phalen, Mary Kay. *Probing the Unknown: The Story of Dr. Florence Sabin*. New York: Thomas Y. Crowell Co., 1969.

Phillips, D. H. "Women in nineteenth-century Wisconsin medicine." *Wisconsin Medical Journal* 71 (1972).

Pirami, E., An eighteenth-century woman physician. *World Medical Journal* 12 (1965): 154.

Power, Eileen. "Some women practitioners of medicine in the Middle
    Ages." *Proceedings of the Royal Society of Medicine* 14 (1921).
Regnault, P., and K. Stephenson. "Dr. Suzanne Noell, the first woman to
    do esthetic surgery." *Plastic and Reconstructive Surgery* 48 (1971):
    133–139.
Robb, Hunter. "Mme. Lachepelle, Midwife." *Bulletin of the Johns Hopkins
    Hospital* 2 (1891).
Robinson, Marion O. *Give My Heart: The Dr. Marian Hilliard Story.*
    Garden City, N.Y.: Doubleday, 1964.
Ross, Ishbel. *Child of Destiny: The Life Story of the First Woman
    Doctor.* New York: Harper Brothers, 1949.
Sanger, Margaret. *Margaret Sanger: An Autobiography.* New York:
    Dover Publications, 1971 (originally 1938).
Shryock, Richard H. "Women in American medicine." In *Medicine in
    America: Historical Essays.* Baltimore: The Johns Hopkins Press,
    1966.
Snively, William D. "Discoverer of the cause of milk sickness [Anna
    Pierce Hobbs Bixby]." *Journal of the American Medical Association*
    196 (1966).
Stern, Madeleine B. *So Much in a Lifetime: The Story of Dr. Isabel
    Burrows.* New York: Messner, 1964.
Strohl, E. Lee. "The fascinating Lady Mary Wortley Montagu, 1689–
    1762." *Archives of Surgery* 89 (1964).
Sturgis, Katharine R. "First woman fellow of the College of Physicians
    of Philadelphia: Memoir of Catharine MacFarlane, 1877–1969."
    *Transactions and Studies of the College of Physicians of Philadelphia*
    38 (1971).
Todd, Margaret. *The Life of Sophia Jex-Blake.* London: Macmillan, 1918.
Traux, Rhoda. *The Doctors Jacobi.* Boston: Little Brown, 1952.
Vietor, Agnecci. *A Woman's Quest: The Life of Marie Zakrzewska, M.D.*
    New York: D. Appleton and Co., 1924.
Wauchope, Gladys Mary. *The Story of A Woman Physician.* Bristol:
    John Wright, 1963.
Wilson, Dorothy Clarke. *Lone Woman: The Story of Elizabeth Blackwell,
    The First Woman Doctor.* Boston: Little Brown, 1970.
————. *Palace of Healing: The Story of Dr. Clara Swain, First Woman
    Missionary Doctor and the Hospital She Founded.* New York:
    McGraw-Hill, 1968.
Yost, Edna. *American Women of Nursing.* Philadelphia: Lippincott,
    1947.

## HISTORY

Alcott, William. "The Young Woman's Book of Health." In *Root of Bitterness*. Edited by N. F. Cott. New York: E. P. Dutton & Co., Inc., 1972 (originally 1855).

Alsop, Gulielma Fell. *History of the Women's Medical College of Philadelphia Pennsylvania. 1850–1950*. Philadelphia: Lippincott, 1950.

*Barker-Benfield, G. J. *The Horrors of the Half-Known Life: Male Attitudes Toward Women and Sexuality in Nineteenth-century America*. New York: Harper and Row, 1976.

Blackwell, Elizabeth. *Address on the Medical Education of Women*. New York: Baptist and Taylor, 1864.

————. "The human element in sex: Being a medical inquiry into the relation of sexual physiology to Christian morality." In *Root of Bitterness*. Edited by N. F. Cott. New York: E. P. Dutton and Co., 1972 (originally 1894).

————. "The influence of women in the profession of medicine." In *Essays in Medical Sociology*. New York: Arno Press and the New York Times, 1972 (originally 1902).

————. *Medicine as a Profession for Women*. New York: Trustees of the New York Infirmary for Women, 1860.

————. *Opening the Medical Profession to Women*. Edited by Mary Roth Walsh. New York: Schocken Books, 1977 (originally 1895).

Blackwell, Emily. "The industrial position of women." *Popular Science Monthly* 23 (July 1883): 388–398.

Blake, John B. "Women and medicine in antebellum America." *Bulletin of the History of Medicine* 39 (1965).

Bolton, H. Carrington. "The early practice of medicine by women." *Popular Science Monthly* 18 (Dec. 1880): 191–201.

Bowditch, Henry I. "The medical education of women." *Boston Medical and Surgical Journal* 105 (August 1881).

Burstyn, Joan N. "Education and sex: The medical case against higher education for women in England, 1870–1900." *Proceedings of the American Philosophical Society* 117 (April 1973): 2.

Chadwick, James Read. "The study and practice of medicine by women." *International Review* (October 1879).

*Daughters of Aesculapius; Stories Written by Alumnae and Students of the Woman's Medical College of Pennsylvania*. Philadelphia: George W. Jacobs, 1897.

Davis, Paulina W. "Female physician." *Boston Medical and Surgical Journal* 41 (1853).

*Ehrenreich, Barbara, and Deirdre English. *Complaints and Disorders: The Sexual Politics of Sickness*. Old Westbury, New York: The Feminist Press, 1973.

\*————. *For Her Own Good: 150 Years of the Experts' Advice to Women.* Garden City, N.Y.: Anchor Press/Doubleday, 1978.

\*————. *Witches, Midwives, and Nurses: A History of Women Healers.* Old Westbury, N.Y.: The Feminist Press, 1973.

Graham, Davis. "The demand for medically educated women." *Journal of the American Medical Association* 6 (1886).

Gregory, George. *Medical Morals . . . and the Importance of Establishing Female Medical Colleges, and Educating and Employing Female Physicians for Their Own Sex.* New York: G. Gregory, 1853.

Haller, John S., Jr. "From maidenhood to menopause: Sex education for women in Victorian America." *Journal of Popular Culture* 6 (Spring 1972).

Hillman, Sara Frazer. *The Founding of Scholarships in the Medical School of the University of Pittsburgh by the Congress of Women's Clubs of Western Pennsylvania, and a Sketch of the Woman Doctor of Yesterday and Today.* 1916.

Hunt, Harriot K. [1805–1875]. "On medical education for women." In *Voices From Women's Liberation.* Edited by Leslie R. Tanner. New York: New American Library, 1970.

Jacobi, Dr. Mary Putnam. "Female invalidism: From a letter to Dr. Edis." In *Root of Bitterness.* Edited by N. F. Cott. New York: E. P. Dutton and Co., 1972 (originally 1895).

Jex-Blake, Sophia. *Medical Women: A Thesis and a History.* New York: Source Books Press, 1970 (originally 1886).

Kinsler, Miriam S. "The American woman dentist: A brief historical review from 1855 through 1968." *Bulletin of the History of Dentistry* 17 (December 1967).

Lander, K. E. "Study of anatomy by women before the nineteenth century." In *Proceedings of the Third International Congress of the History of Medicine.* London, 1972.

McGrew, Elizabeth A. "The history of women in medicine; a symposium: The present. *Bulletin of the Medical Library Association* 44 (1956).

Marshall, Clara. *The Women's Medical College of Pennsylvania: A Historical Outline.* Philadelphia: Blakiston, 1897.

Mead, Kate Hurd. *A History of Women in Medicine from Earliest Times to the Early Nineteenth Century.* Haddam, Conn.: Haddam Press, 1938.

————. "The seven important periods in the evolution of women in medicine." *Bulletin of the Women's Medical College of Pennsylvania* 81 (1931): 6–15.

Munster, L. "Women doctors in medieval Italy." *CIBA Symposium* 10 (1962): 136–140.

Murray, Flora. *Women as Army Surgeons, Being the History of the Women's Hospital Corps in Paris, Wimereux, and Endell Street, Sept. 1914–Oct. 1919.* London: Hodder and Stoughton, 1920.

Nichols, Mary Sargeant Gove. *Lectures to Women on Anatomy and Physiology With an Appendix on Water Cure*. New York: Harper Brothers, 1846.

Smith, Hilda. "Gynecology and ideology in seventeenth-century England." In *Liberating Women's History*. Edited by Berneice A. Carroll. Urbana, Ill.: University of Illinois Press, 1976, pp. 97–114.

Thorne, May. "Women in medicine: The early years." *Postgraduate Medical Journal* 27 (1951).

Verbrugge, M. H. "Women and medicine in nineteenth-century America." *Signs* 1 (Summer 1976).

Waite, Frederick C. *History of the New England Medical College 1848–1874*. Boston, 1950.

Walsh, James J. "Medical education for women." In *Medieval Medicine*. London: A. and C. Black, 1920.

––––––. "Women in medicine." In *History of Medicine in New York: Three Centuries of Medical Progress*. New York: National Americana Society, 1919.

Wood, Ann Douglas. "The fashionable diseases: Women's complaints and their treatment in nineteenth-century America." *Journal of Interdisciplinary History* (Summer 1973).

Wright, Katherine W. "History of women in medicine; a symposium: Nineteenth century or transitional period." *Bulletin of the Medical Library Association* 44 (1956).


GENERAL

Baetjer, Anna M. *Women in Industry: Their Health and Efficiency*. Philadelphia: W. B. Saunders Co., 1946.

*Boston Women's Health Book Collective. *Our Bodies, Ourselves: A Book by and For Women*. 2nd Edition. New York: Simon and Schuster, 1976.

The Diagram Group. *Woman's Body: An Owner's Manual*. New York: Bantam Books, 1977.

*Dreifus, Claudia. *Seizing Our Bodies: The Politics of Women's Health*. New York: Vintage Books, 1978.

Fairfield, L. "Health of professional women." *Medical Woman's Journal* 34 (1927).

*Frankfort, Ellen. *Vaginal Politics*. New York: Quadrangle Books, 1972.

Freeman, Jo., ed. *Women: A Feminist Perspective*. Palo Alto, California: Mayfield Publishing Co., 1975.

Futoran, Jack M., and May Annexton. *Your Body: A Reference Book for Women*. New York: Ballantine Books, 1976.

Hilliard, Marion. *Women and Fatigue: A Woman Doctor's Answer*. Garden City, N.Y.: Doubleday, 1960.

Horos, C. V. *Vaginal Health.* New Canaan, Conn.: Tobey Publishing Co., 1975.

*\*International Journal of Health Services.* "Women and Health: Special Issue." 5 (1975): 167–346.

Lanson, Lucienne. *From Woman to Woman: A Gynecologist Answers Questions About You and Your Body.* New York: Knopf, 1975.

Laversen, N., and S. Whitney. *It's Your Body: A Woman's Guide to Gynecology.* New York: Grosset and Dunlap, 1978.

Lewis, Charles E., and Mary Ann Lewis. "The potential impact of sexual equality on health." *New England Journal of Medicine* (Oct. 1977): 863–869.

Llewellyn-Jones, Derek. *Every Woman and Her Body.* New York: Taplinger, 1971.

McKiever, Margaret I. *The Health of Women Who Work.* Washington, D.C.: U.S. Department of Health, Education, and Welfare, Public Health Service, Government Printing Office, 1965.

Madrigan, F. C. "Are sex mortality differentials biologically caused?" *Milbank Memorial Fund Quarterly* 35 (1957): 202–223.

*\*Milio, Nancy. *The Care of Health in Communities: Access for Outcasts.* New York: Macmillan, 1975.

Orbach, Susie. *Fat is a Feminist Issue.* New York: Pattington Press, 1978.

Parvati, J. *Hygea: A Woman's Herbal.* Berkeley, California: Distributed by Bookpeople, 1978.

Rennie, Susan, and Anna Rubin. "Women's survival catalog: Holistic healing." *Chrysalis* No. 1: 67–69, 1977.

Rush, Anne Kent. *Getting Clear: Body Work for Women.* New York: Random House, 1973.

*Social Policy.* Special Issue on Women and Health. Sept./Oct. 1975.

*\*Stellman, Jean Mage. *Women's Work, Women's Health: Myths and Realities.* New York: Pantheon Books, 1977.

## III.   Bibliographies and Periodicals

*Archives of Sexual Behavior: An Interdisciplinary Research Journal.* New York: Plenum Publishing Co.

Association for the Study of Abortion. *Bibliography Reprint List.* 120 W. 57 St., New York, 10019.

Austin, Helen S., Allison Parelman, and Anne Fisher. *Sex Roles: A Research Bibliography.* Washington, D.C.: U.S. Department of Health, Education, and Welfare, 1976.

*AWIS* [Association for Women in Science] *Newsletter.* Suite 1122, 1346 Connecticut Ave., N.W., Washington, D.C. 20036.

The Barnard College Women's Center. *Women's Work and Women's*

*Studies: A Bibliography.* Published Yearly. Old Westbury, N.Y.: The Feminist Press.

Barnes, D. L. *Rape: A Bibliography 1965–1975.* Troy, N. Y.: Whiteson, Pub., 1977.

Bullough, Vern L., and Barrett Wayne Elcaro. *A Bibliography of Prostitution.* New York: Garland Publishing Co., 1977.

*Canadian Newsletter of Research on Women* [includes all feminist publications, bibliography of current scholarship, etc.]. Dept. of Sociology, Ontario Institute for Studies in Education, 252 Bloor St., W., Toronto, Ontario Canada.

Carey, Emily A. *Women: Sexuality, Psychology and Psychotherapy— A Bibliography.* Womanspace: A Feminist Therapy Collective, 636 Beacon St., Boston, Mass. 02215.

Chaff, S. L. *Women In Medicine: A Bibliography of the Literature on Women Physicians.* Metuchen, N. J.: Scarecrow Press, 1977.

*Davis, Audrey B. *Bibliography on Women: With Special Emphasis on Their Sex Roles in Science and Society.* New York: Science History Publications, 1974.

Glenn, Sara, et al. *Women and Society: Bibliography.* Boston University School of Social Work. Boston, Mass.

*Hughes, Marija Matich. *The Sexual Barrier: Legal, Medical, Economic and Social Aspects of Sex Discrimination* [an annotated bibliography with over 8,000 items]. Washington, D.C. Available from 500 23rd St. N.W. Box B203, Washington, D.C. 20037.

*Hunt, Vilma K. *Women, Work and Health: A Bibliography.* Evanston, Illinois: Program on Women, Northwestern University, 1978.

*Jacobs, Sue-Ellen. *Women in Perspective: A Guide for Cross-Cultural Studies* [a bibliography]. Urbana, Illinois: University of Illinois Press, 1976.

Kemmer, Elizabeth Jane. *Rape and Rape Related Issues. An Annotated Bibliography.* New York: Garland Publishing, 1977.

Key, Mary Ritchie. *Male/Female Language, with a Comprehensive Bibliography.* Metuchen, N. J.: Scarecrow Press, 1975.

Marshall, Joan F., Susan Morris, and Steven Polgar. "Culture and natality: A preliminary classified bibliography." *Current Anthropology* 13 (April 1972): 268–277.

*Psychology of Women Quarterly.* New York: Human Sciences Press.

Rosenberg, Marie B., and Len V. Bergstrom. *Women and Society: A Critical Review of the Literature with a Selected Annotated Bibliography.* Beverly Hills, California: Sage Publications, 1975.

*Ruzek, Sheryl K. *Women and Health Care: An Annotated Bibliography.* Available from Program on Women, Northwestern University, Evanston, Illinois 60201.

St. Louis Feminist Research Project. *The Rape Bibliography: A Collection of Abstracts.* St. Louis: Edy Netter, 1976.

Santa Cruz Women's Health Center. *Annotated Bibliography on Lesbian Health Issues.* Available from 250 Locust St., Santa Cruz, California 95060.

*Sex Roles: A Journal of Research.* New York: Plenum Publishing Co.

*Signs: A Journal of Women in Culture and Society.* Chicago, University of Chicago Press.

Tillman, Randi S. "Women in dentistry—A review of the literature." *Journal of the American Dental Association* 91 (1975): 1214–1215.

*Women and Health: Issues in Women's Health Care* [a journal]. Women and Health, Biological Science Program. SUNY/College at Old Westbury, Old Westbury, N.Y. 11568.

*Women's Studies Abstracts* [issued quarterly]. Palmyra, New York: Rush Publishing Co.